Soil Organic Matter:
Analysis and Interpretation

Related Society Publications

Interactions of Soil Minerals with Natural Organics and Microbes

Soil Fertility and Organic Matter as Critical Components of Production Systems

For information on these titles, please contact the ASA, CSSA, SSSA Headquarters Office; Attn.: Marketing; 677 South Segoe Road; Madison, WI 53711-1086. Telephone: (608) 273-8080. Fax: (608) 273-2021.

Soil Organic Matter: Analysis and Interpretation

Proceedings of a symposium sponsored by Divisions S-4 and S-8 of the Soil Science Society of America in Seattle, Washington, 14 Nov. 1994.

Editors
F.R. Magdoff, M.A. Tabatabai, and E.A. Hanlon, Jr.

Editor-in-Chief SSSA
Jerry M. Bigham

Managing Editor
David M. Kral

Associate Editor
Marian K. Viney

SSSA Special Publication Number 46

Soil Science Society of America, Inc.
Madison, Wisconsin, USA
1996

Cover Design: Patricia Scullion

Copyright © 1996 by the Soil Science Society of America, Inc.

ALL RIGHTS RESERVED UNDER THE U.S. COPYRIGHT ACT OF 1976 (P.L. 94-553)

Any and all uses beyond the limitations of the "fair use" provision of the law require written permission from the publisher(s) and/or the author(s); not applicable to contributions prepared by officers or employees of the U.S. Government as part of their official duties.

Soil Science Society of America, Inc.
677 South Segoe Road, Madison, WI 53711 USA

Library of Congress Registration Number: 96-068997
Printed in the United States of America

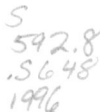

CONTENTS

Foreword vii
Preface ix
Contributors xi
Conversion Factors for SI and non-SI Units xiii

1 Soil Organic Matter Testing: An Overview
   M.A. Tabatabai 1

2 Soil Organic Matter Fractions and Implications
   for Interpreting Organic Matter Tests
   Fred Magdoff 11

3 Estimation of Soil Organic Matter by Weight Loss-On-Ignition
   E.E. Schulte and B.G. Hopkins 21

4 Using Soil Organic Matter to Help Make Fertilizer
   and Pesticide Recommendations
   K.D. Frank and F.W. Roeth 33

5 Assessing Soil Quality by Testing Organic Matter
   Lawrence J. Sikora, Cynthia A. Cambardella,
     Vladimir Yakovchenko, and John W. Doran 41

6 Carbon Fractions in Compost and Compost Maturity Tests
   Charles L. Henry and Robert B. Harrison 51

FOREWORD

Soil organic matter has long been known for its central role in many functions in the soil. They range from controlling nutrient availability to modifying the global carbon budget. In recent and more varieties of organic materials, including municipal sewage sludge, industrial organic wastes, as well as crop residues and animal manures, are returned to the soil organic matter is not always clearly understood. Efforts in optimizing crop production, minimizing environmental pollution, and enhancing soil quality all require a better understanding of the nature of soil organic matter for its proper management. In spite of its importance, there has been a lack of concerted effort in developing appropriate analytical methods for proper characterization and quantification of soil organic matter. As nutrient management and pesticide application recommendations become more precise, demand for more accurate determination of soil organic matter also is increasing. This publication is a timely response to this increasing demand. It provides a state-of-the-art review on the current methods and calls attention to the need for developing specific and improved tests for characterizing soil organic matter. undoubtedly, soil organic matter will continue to provide a challenging field for future research.

H.H. Cheng, President
Soil Science Society of America

PREFACE

Interest in soil organic matter has increased substantially during the last few decades. This has been brought about by a deeper appreciation for organic matter's central role in so many soil processes and properties that are critical for crop growth and environmental quality. The relatively recent activity focused on a better understanding of soil quality has also played a role in the enhanced appreciation for the importance of organic matter.

One area that has not received sufficient attention has been the current and proposed soil organic matter tests and their interpretations. Organic matter tests are presently used by some states to modify soil fertility recommendations. In addition, the labels on a number of herbicides call for modifying application rates based on soil organic matter level. Following a discussion at the 1993 business meeting of the Soil Testing and Plant Analysis Committee (S877), it was decided to request that Divisions S-4 and S-8 jointly sponsor a symposium on the practical issues of soil organic matter testing and test interpretation at the annual meeting to be held in Seattle, WA, in 1994. The two divisions agreed to cosponsor the symposium, which was organized by Fred Magdoff, Chair of S877.

The written versions of talks presented at the symposium are presented in this publication. The first chapter by M.A. Tabatabai covers the procedures of currently used soil organic matter tests. The second chapter by Fred Magdoff discusses potential problems associated with test interpretations. The third chapter by E.E. Schulte and B.G. Hopkins evaluates the weight loss on ignition procedure. The fourth chapter by K.D. Frank and F.W. Roeth presents information on how organic matter is currently used to modify recommendations for lime, fertilizers, and herbicides. The fifth chapter by L. J. Sikora, C. Cambardella, V. Yakovchenko, and J.W. Doran evaluate various proposed tests for assessing changes in soil quality. The sixth chapter by C.L. Henry and R.D. Harrison compares tests proposed for evaluating compost maturity.

These chapters summarize the state of the art and practice of testing for soil organic matter, test interpretation, and using results to modify recommendations for field practices. They should also stimulate researchers and extension specialists to continue to seek improvements in soil organic matter testing-recommendation systems.

Editors
Fred Magdoff
Ed Hanlon
Ali Tabatabai

CONTRIBUTORS

Cynthia A. Cambardella	Soil Scientist, USDA-ARS-MWA, National Soil Tilth Laboratory, 2150 Pammel Drive, Ames, IA 50011-4420
John W. Doran	Soil Scientist, USDA-ARS, 116 Keim Hall, University of Nebraska, Lincoln, NE 68583
K. D. Frank	Extension Agronomist, Department of Agronomy, University of Nebraska, Lincoln, NE 68583-0916
Robert B. Harrison	College of Forest Resources, 234 Bloedel Hall, University of Washington, Seattle, WA 98195-2100
Charles L. Henry	College of Forest Resources, 218 Bloedel Hall, University of Washington, Seattle, WA 98195-2100
Bryan G. Hopkins	Laboratory Director, Servi-Tech Laboratories, P.O. Box 169, Hastings, NE 68902
Fred Magdoff	Professor of Soil Science, Department of Plant and Soil Science, Hills Building, University of Vermont, Burlington, VT 05405
F. W. Roeth	Professor of Agronomy, Department of Agronomy, South Central Research and Extension Center, Box 66, Clay Center, NE 68933
E. E. Schulte	Professor of Soil Science (Emeritus), University of Wisconsin-Madison, 518 S. Owen Drive, Madison, WI 53711
Lawrence J. Sikora	Microbiologist, USDA-ARS, Soil Microbial Systems Laboratory, Building 318, BARC-East, Beltsville, MD 20705
M. A. Tabatabai	Professor of Soil Chemistry, Department of Agronomy, Iowa State University, Ames, IA 50011-1010
Vladimir Yakovchenko	Soil Scientist, USDA-ARS, Soil Microbial Systems Laboratory, Building 318, BARC-East, Beltsville, MD 20705

Conversion Factors for SI and non-SI Units

Conversion Factors for SI and non-SI Units

To convert Column 1 into Column 2, multiply by	Column 1 SI Unit	Column 2 non-SI Units	To convert Column 2 into Column 1, multiply by
Length			
0.621	kilometer, km (10^3 m)	mile, mi	1.609
1.094	meter, m	yard, yd	0.914
3.28	meter, m	foot, ft	0.304
1.0	micrometer, μm (10^{-6} m)	micron, μ	1.0
3.94×10^{-2}	millimeter, mm (10^{-3} m)	inch, in	25.4
10	nanometer, nm (10^{-9} m)	Angstrom, Å	0.1
Area			
2.47	hectare, ha	acre	0.405
247	square kilometer, km^2 (10^3 m)^2	acre	4.05×10^{-3}
0.386	square kilometer, km^2 (10^3 m)^2	square mile, mi^2	2.590
2.47×10^{-4}	square meter, m^2	acre	4.05×10^3
10.76	square meter, m^2	square foot, ft^2	9.29×10^{-2}
1.55×10^{-3}	square millimeter, mm^2 (10^{-3} m)^2	square inch, in^2	645
Volume			
9.73×10^{-3}	cubic meter, m^3	acre-inch	102.8
35.3	cubic meter, m^3	cubic foot, ft^3	2.83×10^{-2}
6.10×10^4	cubic meter, m^3	cubic inch, in^3	1.64×10^{-5}
2.84×10^{-2}	liter, L (10^{-3} m^3)	bushel, bu	35.24
1.057	liter, L (10^{-3} m^3)	quart (liquid), qt	0.946
3.53×10^{-2}	liter, L (10^{-3} m^3)	cubic foot, ft^3	28.3
0.265	liter, L (10^{-3} m^3)	gallon	3.78
33.78	liter, L (10^{-3} m^3)	ounce (fluid), oz	2.96×10^{-2}
2.11	liter, L (10^{-3} m^3)	pint (fluid), pt	0.473

CONVERSION FACTORS FOR SI AND NON-SI UNITS

Mass

To convert Column 1 into Column 2, multiply by	Column 1 SI Unit	Column 2 non-SI Unit	To convert Column 2 into Column 1, multiply by
2.20×10^{-3}	gram, g (10^{-3} kg)	pound, lb	454
3.52×10^{-2}	gram, g (10^{-3} kg)	ounce (avdp), oz	28.4
2.205	kilogram, kg	pound, lb	0.454
0.01	kilogram, kg	quintal (metric), q	100
1.10×10^{-3}	kilogram, kg	ton (2000 lb), ton	907
1.102	megagram, Mg (tonne)	ton (U.S.), ton	0.907
1.102	tonne, t	ton (U.S.), ton	0.907

Yield and Rate

0.893	kilogram per hectare, kg ha^{-1}	pound per acre, lb acre^{-1}	1.12
7.77×10^{-2}	kilogram per cubic meter, kg m^{-3}	pound per bushel, lb bu^{-1}	12.87
1.49×10^{-2}	kilogram per hectare, kg ha^{-1}	bushel per acre, 60 lb	67.19
1.59×10^{-2}	kilogram per hectare, kg ha^{-1}	bushel per acre, 56 lb	62.71
1.86×10^{-2}	kilogram per hectare, kg ha^{-1}	bushel per acre, 48 lb	53.75
0.107	liter per hectare, L ha^{-1}	gallon per acre	9.35
893	tonnes per hectare, t ha^{-1}	pound per acre, lb acre^{-1}	1.12×10^{-3}
893	megagram per hectare, Mg ha^{-1}	pound per acre, lb acre^{-1}	1.12×10^{-3}
0.446	megagram per hectare, Mg ha^{-1}	ton (2000 lb) per acre, ton acre^{-1}	2.24
2.24	meter per second, m s^{-1}	mile per hour	0.447

Specific Surface

10	square meter per kilogram, m^2 kg^{-1}	square centimeter per gram, cm^2 g^{-1}	0.1
1000	square meter per kilogram, m^2 kg^{-1}	square millimeter per gram, mm^2 g^{-1}	0.001

Pressure

9.90	megapascal, MPa (10^6 Pa)	atmosphere	0.101
10	megapascal, MPa (10^6 Pa)	bar	0.1
1.00	megagram, per cubic meter, Mg m^{-3}	gram per cubic centimeter, g cm^{-3}	1.00
2.09×10^{-2}	pascal, Pa	pound per square foot, lb ft^{-2}	47.9
1.45×10^{-4}	pascal, Pa	pound per square inch, lb in^{-2}	6.90×10^3

(continued on next page)

Conversion Factors for SI and non-SI Units

To convert Column 1 into Column 2, multiply by	Column 1 SI Unit	Column 2 non-SI Units	To convert Column 2 into Column 1, multiply by
		Temperature	
$1.00\ (K - 273)$	Kelvin, K	Celsius, °C	$1.00\ (°C + 273)$
$(9/5\ °C) + 32$	Celsius, °C	Fahrenheit, °F	$5/9\ (°F - 32)$
		Energy, Work, Quantity of Heat	
9.52×10^{-4}	joule, J	British thermal unit, Btu	1.05×10^3
0.239	joule, J	calorie, cal	4.19
10^7	joule, J	erg	10^{-7}
0.735	joule, J	foot-pound	1.36
2.387×10^{-5}	joule per square meter, J m^{-2}	calorie per square centimeter (langley)	4.19×10^4
10^5	newton, N	dyne	10^{-5}
1.43×10^{-3}	watt per square meter, W m^{-2}	calorie per square centimeter minute (irradiance), cal cm^{-2} min^{-1}	698
		Transpiration and Photosynthesis	
3.60×10^{-2}	milligram per square meter second, mg m^{-2} s^{-1}	gram per square decimeter hour, g dm^{-2} h^{-1}	27.8
5.56×10^{-3}	milligram (H$_2$O) per square meter second, mg m^{-2} s^{-1}	micromole (H$_2$O) per square centimeter second, μmol cm^{-2} s^{-1}	180
10^{-4}	milligram per square meter second, mg m^{-2} s^{-1}	milligram per square centimeter second, mg cm^{-2} s^{-1}	10^4
35.97	milligram per square meter second, mg m^{-2} s^{-1}	milligram per square decimeter hour, mg dm^{-2} h^{-1}	2.78×10^{-2}
		Plane Angle	
57.3	radian, rad	degrees (angle), °	1.75×10^{-2}

CONVERSION FACTORS FOR SI AND NON-SI UNITS

Electrical Conductivity, Electricity, and Magnetism

To convert Column 1 into Column 2, multiply by	Column 1 SI Unit	Column 2 non-SI Unit	To convert Column 2 into Column 1, multiply by
10	siemen per meter, S m^{-1}	millimho per centimeter, mmho cm^{-1}	0.1
10^4	tesla, T	gauss, G	10^{-4}

Water Measurement

9.73 × 10^{-3}	cubic meter, m^3	acre-inches, acre-in	102.8
9.81 × 10^{-3}	cubic meter per hour, m^3 h^{-1}	cubic feet per second, ft^3 s^{-1}	101.9
4.40	cubic meter per hour, m^3 h^{-1}	U.S. gallons per minute, gal min^{-1}	0.227
8.11	hectare-meters, ha-m	acre-feet, acre-ft	0.123
97.28	hectare-meters, ha-m	acre-inches, acre-in	1.03 × 10^{-2}
8.1 × 10^{-2}	hectare-centimeters, ha-cm	acre-feet, acre-ft	12.33

Concentrations

1	centimole per kilogram, cmol kg^{-1}	milliequivalents per 100 grams, meq 100 g^{-1}	1
0.1	gram per kilogram, g kg^{-1}	percent, %	10
1	milligram per kilogram, mg kg^{-1}	parts per million, ppm	1

Radioactivity

2.7 × 10^{-11}	becquerel, Bq	curie, Ci	3.7 × 10^{10}
2.7 × 10^{-2}	becquerel per kilogram, Bq kg^{-1}	picocurie per gram, pCi g^{-1}	37
100	gray, Gy (absorbed dose)	rad, rd	0.01
100	sievert, Sv (equivalent dose)	rem (roentgen equivalent man)	0.01

Plant Nutrient Conversion

	Elemental	Oxide	
2.29	P	P$_2$O$_5$	0.437
1.20	K	K$_2$O	0.830
1.39	Ca	CaO	0.715
1.66	Mg	MgO	0.602

1 Soil Organic Matter Testing: An Overview

M. A. Tabatabai

Iowa State University
Ames, Iowa

Soil organic matter is defined as the organic fraction of the soil exclusion of undecayed plant and animal residues. Often used synonymously with humus (SSSA, 1987). Organic matter is the most complex, dynamic, and reactive soil component. It is an important soil constituent, because it contributes to plant growth and development through its effect on the chemical, biological, and physical properties of soils. It has a nutritional function in that it serves as a source of N, P, and S for plant growth; a biological function in that it profoundly affects the activities of microflora and microfaunal organisms; and a physical function that it promotes good soil structure, thereby improving tilth, aeration, and water retention (Stevenson, 1994). In addition, organic matter in soils is involved in cementing soil particles into structural units (aggregates), chelating metals, providing buffer action, contributing to the cation-exchange capacity, and mineralization of important plant nutrients, and affecting the bioactivity, persistence, and biodegradability of pesticides. Because of the importance of organic matter in soils, its estimation is important in disciplines ranging from soil fertility, chemistry, and physics to land planning and soil productivity.

A variety of methods are available for determination of organic matter in soils. Early work involved determination of the changes in weight of a soil sample resulting from destruction of organic compounds with H_2O_2 or ignition at high temperature, up to 850°C for 30 min (Rather, 1917; Robinson, 1927; Degtjareff, 1930; Ball, 1964). Both techniques are subject to error. The H_2O_2 method proposed by Robinson (1927) has serious limitations in that the oxidation of organic matter by this reagent is incomplete, and the extent of oxidation varies among soils. This method is therefore unsatisfactory as a means of determining total organic matter of soils, but it can be useful as means of comparing the readily oxidizable material in different soils (Broadbent, 1965). Additional errors are introduced both in filtering and drying the oxidized residue and the filtrate at 110°C.

The ignition method overestimates organic matter in soils because both inorganic, mainly the hydrated aluminosilicates, and organic constituents lose weight during heating; however, if the soil is pretreated with a mixture of HCl and HF to remove the hydrated minerals, loss on ignition then gives a valid estimate of organic matter (Rather, 1917). Soon and Abboud (1991) compared the results

---

Copyright © 1996. Soil Science Society of America, 677 S. Segoe Rd., Madison, WI 53711, USA.
Soil Organic Matter: Analysis and Interpretation. SSSA Special Publication no. 46.

obtained by several methods used for determination of organic C in soils and concluded that the loss-on-ignition procedure, even when allowance was made for clay content, was the least satisfactory of the methods tested.

Organic matter in soils is usually estimated by multiplying the organic C concentration by a constant factor. The factor most commonly used is 1.724. This factor was proposed by Sprengel in 1826 (C.A. Black, 1964, personal communication) assuming that soil organic matter is made up of 58% C. Since then several workers have shown that this conversion factor for surface soils can vary from 1.724 to 2.00 and for subsurface soil is about 2.5 (Broadbent, 1965). The appropriate factor must be determined experimentally for each soil; however, neither the direct determination of organic matter nor the calculation of organic matter content of soils is completely accurate. Therefore, as recommended by Nelson and Sommers (1982), because of the problems associated with determination of organic matter in soils, it is recommended that investigators determine and report organic C concentration as a measure of organic matter in soils.

Several techniques are available for determination of organic C in soils. Each technique has advantages and disadvantages, some of which are labor intensive and others require expensive equipment. This review will cover the most commonly used methods for determination of organic C in soils.

METHODS OF DETERMINING ORGANIC CARBON

Difference between Total Carbon and Inorganic Carbon

Organic C can be determined from the difference between total C and inorganic C (i.e., total C and inorganic C are determined on separate samples: organic C = total C − inorganic C). A number of methods are available for determination of total C in soils. With most of those methods, the results obtained for noncalcareous soils are equal to organic C. When estimation of organic C in a calcareous soil is desired, it is calculated from the difference between total C and inorganic C determined independently. Comparison of the methods used for determination of total C in soils is shown in Table 1–1. These methods range from dry combustion by using a resistance or induction furnace or by automated methods to wet combustion involving combustion train or Van Slyke-Neil apparatus.

In using a resistance furnace, the sample is mixed with CuO and heated to about 1000°C in a stream of O_2 to convert all C in the sample to CO_2. In the use of induction furnace, the sample is mixed with Fe and accelerators (Sn and Cu) and rapidly heated to about 1650°C in a stream of O_2 to convert all C in sample to CO_2. In both methods, the CO_2 evolved is determined by gravimetric or titrimetric methods.

In dry combustion automated methods, the sample is mixed with catalysts or accelerators and heated in resistance or induction furnace in a stream of O_2 to convert all C in the sample to CO_2. The CO_2 thus evolved is determined by gas chromatographic methods (thermal conductivity or flame ionization detector), gravimetrically, or titrimetrically.

Table 1–1. Methods used for determination of total C in soils.†

Method	Principles	CO_2 determination
Dry combustion		
Resistance furnace	Sample is mixed with CuO and heated to about 1000°C in a stream of O_2 to convert all C in sample to CO_2	Gravimetric Titrimetric
Induction furnace	Sample is mixed with Fe, Cu, and Sn and rapidly heated to >1650°C in a stream of O_2 to convert all C in sample to CO_2	Gravimetric Titrimetric
Automated	Sample is mixed with catalysts or accelerators and heated with resistance or induction furnaces in a stream O_2 to convert all C in sample to CO_2	Gas chromatography Gravimetric Conductimetric
Wet combustion		
Combustion train	Sample is heated with $K_2Cr_2O_7$–H_2SO_4–H_3PO_4 mixture in a CO_2-free air stream to convert all C in sample to CO_2	Gravimetric Titrimetric
Van Slyke-Neil apparatus	Sample is heated with $K_2Cr_2O_7$–H_2SO_4–H_3PO_4 mixture in a combustion tube attached to a Van Slyke-Neil apparatus to convert all C in sample to CO_2.	Manometric

† Nelson and Sommers (1982).

A wet combustion method involving a combustion train is also available (Allison, 1960). In this method, the sample is heated with a mixture containing $K_2Cr_2O_7$, H_2SO_4, and H_3PO_4 in a CO_2-free air stream to convert all C in the sample to CO_2. The evolved CO_2 is determined gravimetrically or titrimetrically. The same mixture can be used to oxidize total C in a soil sample to CO_2 in a Van Slyke-Neil apparatus, and the CO_2 evolved is measured manometrically. Although the methods described above give accurate results with high precision, the disadvantages are that some of the methods are time-consuming and require a leakfree O_2 sweep train (e.g., dry combustion with resistance furnace or wet combustion with combustion train) or require an expensive furnace (e.g., induction furnace). The operation of the Van Slyke-Neil apparatus requires great skill, and the equipment is somewhat expensive and easily damaged.

Several methods are available for determination of inorganic C in soils. These include (i) neutralization of the carbonate with acid and back titration of the excess acid; (ii) determination of Ca and Mg in an acid leachate; and (iii) dissolution of carbonate in acid and determination of the CO_2 by measuring the volume or pressure of the CO_2 or by titrimetry, sample weight loss, infrared spectrometry, gas chromatography, and thermogravimetry (Nelson & Sommers, 1982).

The main disadvantage of the approach described above for determination of organic C is that two separate analyses are required. Total C determination requires special equipment. In addition, organic C calculated by difference has some error. Therefore, the results obtained by this approach are neither accurate nor precise.

Determination of Total Carbon after Removal of Inorganic Carbon

Alternately, organic C in soils can be determined after removal of the inorganic C with acid pretreatment (i.e., organic C = total C; Table 1–2). This technique gives accurate results if dolomite is absent from soil; not all dolomite in soils can be removed by acid treatment. In addition, this technique requires special equipment (Allison, 1960).

Dichromate Oxidation

Early work by Rogers and Rogers (1848) showed that a dichromate–sulfuric acid mixture can be used for wet oxidation of organic substances. Later Warrington and Peake (1880) and Cameron and Breazeale (1904) unsuccessfully applied the method to determination of organic C in soils; according to Clark and Ogg (1942) their results were low. A decade later, Ames and Gaither (1914) were the first to show that complete oxidation of total C in soils can be accomplished by a dichromate–sulfuric acid mixture. The present methods fall into two groups (Table 1–2). In one group, the dichromate oxidation is employed without external heat. In the second group of methods, the dichromate oxidation is employed with external heat. In both groups of methods, the remaining $Cr_2O_7^{2-}$ is back titrated with ferrous ammonium sulfate by using one of three oxidation-reduction indicators (diphenylamine, o-phenanthroline, or N-pheylanthranilic acid). The reactions involved are as follows:

$$2\ Cr_2O_7^{2-} + 3\ C^o + 16\ H^+ = 4\ Cr^{3+} + 3\ CO_2 + 8\ H_2O \qquad [1]$$

$$Cr_2O_7^{2-} + 6\ Fe^{2+} + 14\ H^+ = 2\ Cr^{3+} + 6\ Fe^{3+} + 7\ H_2O. \qquad [2]$$

This approach was first introduced by Schollenberger (1927), and since then a number of modifications have been proposed (Nelson & Sommers, 1982). Most of these modifications have been related to the type and concentration of the acid used and whether or not external heat is employed.

Table 1–2. Methods used for determination of organic C in soils.†

Method	Principles
Difference between total C inorganic C	Total C and inorganic C are determined on separate samples: Organic C = Total C – Inorganic C
Determined as total C after removal of inorganic C	Total C is determined on the sample after removal of inorganic C with an acid pretreatment: Organic C = Total C.
Dichromate oxidation without external heat	Dichromate oxidizes organic C to CO_2 in acid without external heat medium. Amount of $Cr_2O_7^{2-}$ reduced is quantitatively related to organic C present. Not all organic C in sample is oxidized when external heat is omitted, and a correction factors is required.
Dichromate oxidation with external heat	Same as the dichromate method described above but all organic C in the sample is oxidized; no correction factor is needed.

† Nelson and Sommers (1980).

Appreciable concentration of Fe^{2+} may be present in soil under highly reduced conditions, and errors may results when dichromate methods are employed in determination of organic C in such soil samples before drying (see Reaction [2]). Another source of positive error in these methods is the presence of Cl^-. Chloride reacts with dichromate producing chromyl chloride as follows:

$$K_2Cr_2O_7 + 4\ KCl + 6\ H_2O = CrO_2Cl_2 + 6\ KHSO_4 + 3\ H_2O \quad [3]$$

This error can be large in salt affected soils, and appropriate controls (e.g., ignited sample) should be included to account for the amount of dichromate consumed in its reaction with chloride.

The higher oxides of Mn (largely MnO_2) compete with dichromate for oxidizable substances when heated in an acid medium resulting in a negative error:

$$2\ MnO_2 + C^o + 8\ H^+ = CO_2 + Mn^{2+} + 4\ H_2O. \quad [4]$$

The interference from MnO_2 is not a serious error in most soils, because only a small amount of MnO_2 is present in soils. Even in highly manganiferous soils only small fraction of MnO_2 compete with $Cr_2O_7^{2-}$ for oxidation of organic C (Nelson & Sommers, 1982).

Dichromate Oxidation Without External Heat

Rapid titration methods such as those proposed by Schollenberger (1927), Walkley and Black (1934), and Tiurin (1935) are based on the oxidation of organic C in soil with a dichromate–sulfuric acid mixture without an external heat source to accelerate and complete the oxidation (Table 1–2). It is well known that all these methods do not give complete oxidation of organic C, and therefore require a correction factor for calculating the results. The use of a correction factor may introduce an error, because the organic matter in different soils, or even in different horizons of a soil profile, is not always oxidized to the same degree. The correction factor may vary markedly among surface soils, reported values range from 1.19 to 1.40 (Nelson & Sommers, 1982; Soon & Abboud, 1991), and are greater for subsurface than for surface soils.

Dichromate Oxidation with External Heat

The use of external heat to complete the oxidation of organic C by the dichromate–sulfuric acid procedure and a reflux condenser to minimize loss of CrO_2Cl_2 produced from the reaction of $K_2Cr_2O_7$ and Cl_2 in some soils, especially saline soils, was introduced by Mebius (1960). Since then, this method has been widely used. Unlike the methods that are based on the internal heat generated from mixing concentrated sulfuric acid with a dichromate solution added to the soil sample, the Mebius procedure (1960) involves boiling the soil–dichromate–sulfuric acid mixture for 30 min in an Erlenmeyer flask connected to a reflux condenser. The dichromate remaining is back titrated with ferrous ammonium sulfate in the presence of N-phenylanthranilic acid as an indicator. In this method, two blanks are included; one for standardization of the ferrous ammonium sulfate (because it is not stable) and the other a boiled blank to account for self decomposition of the dichromate under boiling conditions. This method was modified by Nelson and Sommers (1975) by performing the digestion for 30 min in a 50-mL Folin-Wu nonprotein N tube placed in an aluminum heating block at 150°C. Another modification was introduced by Yeomans and Bremner (1988) by digesting the sample with $K_2Cr_2O_7$–H_2SO_4 mixture for 30 min in a Pyrex digestion tube in a 40-tube block digester preheated at 170°C. Soon and Abboud (1991) heated the mixture to 135°C. In all these modifications, the excess $K_2Cr_2O_7$ is back titrated with ferrous ammonium sulfate in the presence of N-pheylanthranilic acid as an indicator.

Colorimetric Methods after Oxidation With Dichromate

The wet combustion methods involving dichromate for determination of organic C in soils, described above, have been modified to use colorimetric analysis instead of titration (Carolan, 1948; Graham, 1948; Perrier & Kellog, 1960; Sims & Haby, 1971). Two approaches have been used in these colorimetric methods: (i) colorimetric determination of the amount of unreacted dichromate after the oxidation reaction was completed; the color changes from orange to green, and (ii) measurement of the absorbance of the color complex (violet color) produced from the reaction of Cr^{3+} produced with s-diphenylcarbazide at 540 nm. The procedure in (i) is described by Graham (1948), Carolan (1948), and Sims and Haby (1971). This procedure is based on the fact that oxidation of organic C compounds by dichromate in acid media involves the half reaction: $Cr_2O_7^{2-} + 14\ H^+ + 6e^- = 2\ Cr^{3+} + 7\ H_2O$. Because the appearance of Cr^{3+} in the reaction medium of dichromate oxidation of organic C is proportional to the quantity of organic C oxidized, analysis of the solution for Cr^{3+} should provide an index of organic C content of the sample oxidized. Chromium (III) exists in acid aqueous solution as a stable hexaquo ion $[Cr(H_2O)_6]^{3+}$ (Cotton & Wilkinson, 1962). But, Cr(III) has two broad maxima in the visible range, one near 450 nm and the other near 600 nm. The dichromate ion also has an absorption maximum near 450 nm but does not absorb near 600 nm. Therefore, Sims and Haby (1971) showed that oxidation of organic C with dichromate followed by the measurement of the absorbance of Cr (III) near 600 nm should provide a satisfactory routine method for determination of organic C in soils. They also showed that the organic C values obtained by this method for 87 soil samples were significantly correlated ($r = 0.986$) with those obtained by the titration method described by Walkley (1947). As with all colorimetric methods, the solution should be free of turbidity.

Automated Instruments

Organic C in noncalcareous soils can be determined by dry-combustion techniques. For calcareous soils, the concentration of the inorganic C must be subtracted from the total C values in calculation of organic C (see Difference between Total Carbon and Inorganic Carbon). Dry-combustion techniques are not used for determination of total or organic C of soils, because until relatively recently the techniques were manual, complicated, and time consuming compared with wet combustion methods. Most of these problems have been overcome by development of automated combustion instruments.

Several automated combustion instruments have been evaluated for determination of total C in soils. The Laboratory Equipment Corporation of St. Joseph, MI, supplies five automated C analyzers. These include the Leco 70-second instrument, the Leco IR-12 Carbon determinator, the Leco DC-12 Duo-Carbon Analyzer, the Leco CS-46 Simultaneous Carbon/Sulfur Determinator, and the Leco CHN-600 and CHN-800 Analyzers. The DC-12 Duo-Carb instrument permits the determination of total C and organic C, the CS-46 instrument allows the determination of both total C and total S, and the CHN-600 and CHN-800 instruments allow the simultaneous determination of total C, H, and S. Evaluation of the Leco CR-12 determinator and the Leco CNH-600 Elemental Analyzer has shown that the former is satisfactory for routine determination of total C in soils, and that the latter gives accurate results for simultaneous determination of total C and N in soils (McGeehan & Naylor, 1988; Yeomans & Bremner, 1991). Another automated combustion instrument now available for the simultaneous determination of total C and total S is the I. R.-Matic C-S VK-111 AS Analyzer supplied by Kukusai Electric Co., Japan. The principles involved in operation of these instruments and their application to determination of total or organic C in soils are described by Tabatabai and Bremner (1991). In addition to the instruments described above, two other instruments are commercially available. These instruments involve an noncatalyzed, flash-combustion of the sample in a quartz tube in an O_2–He atmosphere. One is the instrument developed by Hercules, Wilmington, DE, for the simultaneous determination of C, H, N, and S, and the other is the Model 1106 Elemental Analyzer developed by Carlo Erba Strumentazione, Milan, Italy, for determination of C, H, N, O, and S. These last two instruments, however, have not been evaluated for determination of C, N, and S in soils, but because of the small sample size required, it seems unlikely to prove useful for soil analysis.

SUMMARY

Soil organic matter is one of the most important properties affecting chemical reactions and the availability of nutrient elements to plants. Chemically, it is a complex mixture of plant, animal, and microbial residues, fresh and at all stages of decomposition, and the relatively resistant soil humus. It is difficult to estimate the amount of organic matter in soils quantitatively. The procedures are based on determination of changes in weight of a soil sample resulting from destruction of organic compounds by H_2O_2 treatment or ignition at high temperature. Both methods are subject to errors. Therefore, organic C concentration is determined as a measure of organic matter in soils. Wet oxidation methods are available that give accurate and precise results provided that external heat is applied for boiling the soil sample and potassium dichromate in an acid medium. When the wet oxidation methods are used without external heat, a correction factor must be used, which varies among soils.

Automated instruments are commercially available that can be used for determination of total C in soils, but a correction must be made for inorganic C present in calcareous soils. For acid soils, the results obtained by automated instruments are in agreement with those obtained by wet oxidation methods involving external heat. Each of the methods described have advantages and limitations, and the method selected should be carefully assessed in terms of the purpose of analyzing soils for organic C or organic matter.

REFERENCES

Ames, W.J., and E.W. Gaither. 1914. Determination of carbon in soils and soil extracts. J. Ind. Eng. Chem. 6:561.

Allison, L.E. 1960. Wet-combustion apparatus for organic and inorganic carbon in soils. Soil Sci. Soc. Am. Proc. 24:36–40.

Ball, D.F. 1964. Loss-on-ignition as an estimate of organic matter and organic carbon in non- calcareous soils. J. Soil Sci. 15:84–92.

Broadbent, F.E. 1965. Organic matter. p. 1397–1400. In C.A. Black (ed.) Methods of soil analysis. Agron. Monogr. 9. ASA, CSSA, and SSSA, Agron., Madison, WI.

Cameron, F.K., and J.F. Breazeale. 1904. The organic matter in soils and subsoils. J. Am. Chem Soc. 26:29–45.

Carolan, R. 1948. Modification of Graham's method for determining soil organic matter by colorimetric analysis. Soil Sci. 66:241–247.

Clark, N.A., and C.J. Ogg. 1942. A wet combustion method for determining total carbon in soils. Soil Sci. 53:27–35.

Cotton, F.A., and G. Wilkinson. 1962. Advanced inorganic chemistry. John Wiley & Sons, New York.

Degtjareff, W.T. 1930. Determining soil organic matter by means of hydrogen peroxide and chromic acid. Soil Sci. 29:239–245.

Graham, E.R. 1948. Determination of soil organic matter by means of photoelectric colorimeter. Soil Sci. 65:181–183.

Mebius, L.J. 1960. A modified method for the determination of organic carbon in soil. Anal. Chim. Acta 22:120–124.

McGeehan, S.L., and D.V. Naylor. 1988. Automated instrumental analysis of carbon and nitrogen in plant and soil samples. Commun. Soil Sci. Plant Anal. 19:493–505.

Nelson, D.W., and L.E. Sommers. 1975. A rapid and accurate procedure for estimation of organic carbon in soils. Proc. Indiana Acad. Sci. 84:456–462.

Nelson, D.W., and L.E. Sommers. 1982. Total carbon, organic carbon, and organic matter. p. 539–579. In A.L. Page et al. (ed.) Methods of soil analysis. 2nd ed. Part 2. Agron. Monogr. 9. ASA and SSSA, Madison, WI.

Perrier, E.R., and M. Kellog. 1960. Colorimetric determination of soil organic matter. Soil Sci. 90:104–106.

Rather, J.B. 1917. An accurate loss on ignition method for determination of organic matter in soils. Arkansas Agric. Exp. Stn. Bull 140.

Robinson, W.O. 1927. The determination of organic matter in soils by means of hydrogen peroxide. J. Agric. Res. 34:339–356.

Rogers, R.E., and W.R. Rogers. 1848. New method of determining the carbon in native and artificial graphite, etc. Am. J. Sci. (2)5:352.

Schollenberger, C.J. 1927. A rapid approximate method for determining soil organic matter. Soil Sci. 24:65–68.

Sims, J.R., and V.A. Haby. 1971. Simplified colorimetric determination of soil organic matter. Soil Sci. 112:137–141.

Soon, Y.K., and S. Abboud. 1991. A comparison of some methods for soil organic carbon determination. Commun. Soil Sci. Plant Anal. 22:947–954.

Soil Science Society of America. 1987. Glossary of soil science terms. SSSA, Madison, WI.

Stevenson, F.J. 1994. Humus chemistry: Genesis, composition, reactions. 2nd ed. John Wiley & Sons, New York.

Tabatabai, M.A., and J.M. Bremner. 1991. Automated instruments for determination of total carbon, nitrogen, and sulfur in soils by combustion techniques. p. 261–286. *In* K.A. Smith (ed.) Soil analysis: Modern instrumental techniques. 2nd ed. Dekker, New York.

Tiurin, I.V. 1935. Comparative study of the methods for the determination of organic carbon in soils and water extracts of soils. Dokuchaiv Soil Inst. Stud. Genesis Geogr. Soils. 1935:139–158.

Walkley, A. 1947. A critical examination of a rapid method for determining organic carbon in soils: Effect of variations in digestion conditions and inorganic soil constituents. Soil Sci. 63:251–264.

Walkley, A., and I.A. Black. 1934. An examination of the Degtjareff method for determining soil organic matter, and a proposed modification of the chromic acid titration method. Soil Sci. 37:29–38.

Warrington, R., and W.A. Peake. 1880. On the determination of carbon in soils. J. Chem. Soc. (London) 37:617–625.

Yeomans, J.C., and J.M. Bremner. 1988. A rapid and precise method for routine determination of organic carbon in soil. Commun. Soil Sci. Plant Anal. 19:1467–1476.

Yeomans, J.C., and J.M. Bremner. 1991. Carbon and nitrogen analysis of soils by automated combustion techniques. Commun. Soil Sci. Plant Anal. 22:843–850.

2 Soil Organic Matter Fractions and Implications for Interpreting Organic Matter Tests

Fred Magdoff

University of Vermont
Burlington, Vermont

Interest in soil organic matter (SOM) has increased dramatically during the last decade. Many soil testing laboratories are now offering organic matter tests as part of their routine suite of analyses, and most other laboratories will do SOM tests on request. The SOM tests currently being used for routine analysis are the Walkley–Black (WB) and weight loss on ignition (WLOI) procedures. Presently, various laboratories, custom applicators, or end users employ SOM levels to modify N fertilizer, lime, and pesticide recommendations. This chapter will discuss some of the potential pitfalls of commonly used interpretations for these various tests. Other proposed tests or approaches to SOM testing also will be discussed.

SOIL ORGANIC MATTER FRACTIONS

Soil organic matter, in its broadest definition, consists of a number of fractions that are different from one another. The quantities and compositions of the various types of organic matter can have distinct agronomic effects. One simple, but useful classification divides SOM into three fractions: the active, the well-decomposed, and the living organisms. The *active fraction* consists mainly of relatively fresh residues that, when intimately mixed with the soil, are readily mineralized, and provide much of the energy for soil organisms and available N for plant growth during the current growing season. When fresh residues are on the surface in reduced tillage systems this fraction is not as readily available to soil microorganisms because the lack of a favorable environment commonly slows decomposition. *Well-decomposed* humic materials are the source of much of organic matter's cation-exchange capacity (CEC), as well as other adsorption and chelating abilities. The *living organisms* carry out many functions relating to soil structure, fertility, and plant pests. The activity and diversity of soil organisms is critical for nutrient cycling, controlling populations of many pests, promoting

---

Copyright © 1996. Soil Science Society of America, 677 S. Segoe Rd., Madison, WI 53711, USA.
Soil Organic Matter: Analysis and Interpretation. SSSA Special Publication no. 46.

porosity by burrowing activities [e.g., earthworm (*Lumbricus terrestris*) channels], and producing plant hormones, as well as substances that promote soil aggregation.

These three fractions are, of course, interrelated in a number of ways. Some of the living portion of SOM also belongs in the active pool as grazers and predators, going about their normal business, excrete products from organisms they have consumed. Reasonably good supplies of active organic matter are necessary to supply a diverse and thriving soil biological community. As organisms use active organic matter, and mineralize nutrients such as N, the amount of this fraction decreases; however, as mineralization is occurring, a portion substrate C ends up as part of the well-decomposed humic fraction.

KEY SOIL ORGANIC MATTER FRACTIONS AND CROPPING RECOMMENDATIONS

The main uses for SOM tests in agricultural production decision making are for modification of soil fertility (usually N) and pesticide recommendations (see Frank & Roeth, 1996, this publication). For the purposes of using a soil test to estimate soil fertility effects (N availability during the growing season), the test needs to reflect or be correlated with the amount of active SOM. During decomposition of the active fraction, in addition to the liberation of available N (and other nutrients), many of the sticky substances that promote soil aggregation (polysaccharides, polyuronides, as well as fungal hyphae) are produced, enhancing soil physical properties. The amount of well decomposed residues, however, also is important, because it is the source of most of the CEC due to organic matter, and SOM provides a major portion of the CEC for most coarse to medium texture soils (Magdoff & Bartlett, 1985).

The well decomposed fraction is believed to be the most important influence of SOM on pesticide tie-up, and therefore on application rates needed to control pests (Weber & Peters, 1982). The degree of adsorption of pesticides is related to the chemical characteristics of the particular compound as well as soil characteristics. The CEC of organic matter adsorbs compounds that are commonly cations, such as the quaternary N pesticides [e.g., diquat (6,7-dihydrodipyrido(1,2-α:2',1'-c) pyrazinediium ion) and paraquat (1,1'-dimethyl-4,4'-bipyridinium)]. Other basic compounds such as atrazine [(6-chloro-N-ethyl-methyl-N'-(1-methylethyl)-1,3,5-triazine-2,4-diamine)] tend to be protonated and become cations and are better adsorbed at lower pHs in soils. These types of compounds also are adsorbed by CEC on clay and may be held by both coulombic and physical forces (Weber, 1994). The adsorption by soils of carboxylic acid herbicides tends to be influenced by soil pH and organic matter. At low soil pH, these compounds are not ionized and tend to be more strongly adsorbed by lipophilic organic matter groups than when present as anions (Weber, 1994). Herbicides with certain reactive groups such as glyphosate with phosphate and DSMA with arsenate also are adsorbed strongly in soils, although the mineral fraction may be more responsible than organic matter.

POTENTIAL PROBLEMS WITH CURRENT SOIL ORGANIC MATTER TEST INTERPRETATIONS

Estimating Fertility Effects from Soil Organic Matter Tests

There are a number of limitations to using tests that estimate total SOM to predict the soil's current year contribution to plant nutrition (usually available N). The organic matter content of a particular soil is a function of the additions and losses that have occurred with time. The change in SOM during the year, ΔSOM, is

$$\Delta\text{SOM} = \text{additions} - \text{losses} \qquad [1]$$

When SOM is at equilibrium, and there is no net increase or decrease with time, ΔSOM = 0 and Eq. [1] can be rewritten as

$$\text{additions} = \text{losses} \qquad [2]$$

Losses are the amounts of organic matter mineralized annually. They can be expressed as the fraction (k) of the organic matter that mineralizes annually multiplied by the amount present, or k(SOM).

Thus, given conditions of equilibrium, the amount of organic matter that must be added annually to the slowly decomposing pool can be expressed as:

$$\text{additions} = \text{losses} = k(\text{SOM}) \qquad [3]$$

Soils in a given climatic region that are similar in texture and drainage class, will probably have similar k values, and different levels of SOM mainly reflect varying rates of organic residue return over time as well as differences in rotations and frequency of tillage. In this situation, differences in SOM may be correlated with microbial biomass and activity (Schnurer et al., 1985) and SOM content also may be correlated with nutrient availability from mineralization. Because k is the same, k(SOM) varies directly with SOM. A hypothetical example for a well-drained soil low in organic matter and a $k = 0.03$ that was amended with different amounts of manure shows a rapid increase in N mineralized as SOM increases (Fig. 2–1).

When soils of very different properties are compared, there is a complicating factor: as SOM increases there is a tendency for the rate of mineralization, k, to decrease (Magdoff, 1978, 1991). This occurs because SOM content is highly dependent on the rate of SOM loss (or mineralization) when climate and agricultural practices are similar. And the rate of SOM mineralization is strongly influenced by soil properties such as texture and internal drainage.

As soil clay content increases, the amount of soil organic matter tends to increase as well (Fig. 2–2). The formation of organo-mineral complexes helps to stabilize SOM as does the formation of aggregates. This stabilization is probably a physical sequestering of materials out of easy access to soil organisms, and may provide some protection from extra-cellular enzymatic attack. Thus, soils higher

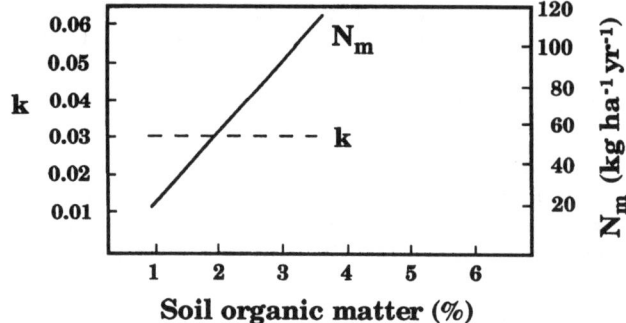

Fig. 2–1. Hypothetical relationship between soil organic matter (SOM) and amount of N mineralized (N_m) for soils with the same decomposition rate (k), but different past histories of manure applications.

in clay and silt content tend to have higher levels of organic matter; however, while there is an organic matter enrichment of the clay and silt fractions relative to the whole soil, as clay content increases, there is a decrease in the extent of SOM enrichment of the clay fraction (Christensen, 1992). Higher clay and silt content tends to be associated with lower rates of SOM decomposition under cropping than in coarser texture soils (Dalal & Mayer, 1986). There also is a tendency for SOM to be higher in soils with restricted drainage than in well-drained and excessively-drained soils. This is mainly due to the lack of aerobic conditions under restricted drainage resulting in a low rate of SOM decomposition.

Given conditions of decreasing *k* as SOM increases, the amount of N mineralized tends to be relatively low at both low and high SOM (Fig. 2–3). One state (Missouri) that uses SOM levels to modify N recommendations implicitly takes this phenomena into account and decreases the N credit given per unit of SOM as SOM increases.

There are a number of other phenomena that make it difficult to use SOM levels as indicators of potential N supply. Organic molecules are added to the

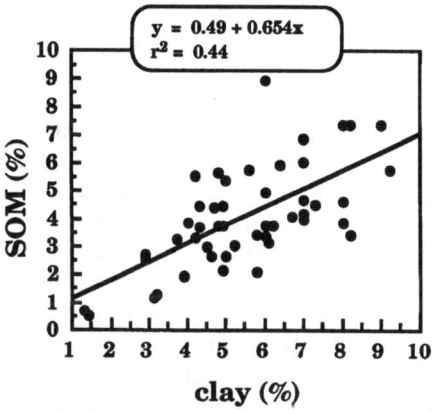

Fig. 2–2. Soil organic matter (SOM) vs. clay content (Magdoff, 1975–1994, unpublished data).

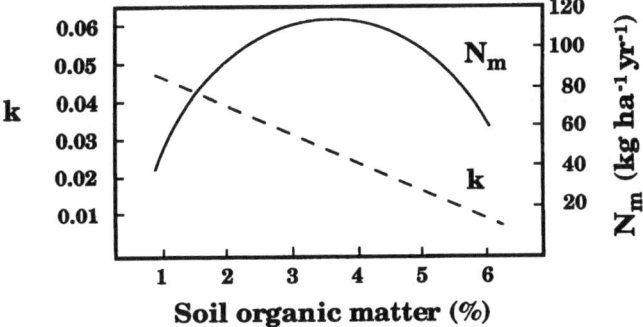

Fig. 2–3. Hypothetical relationship between soil organic matter (SOM) and amount of N mineralized (N_m) for soils with very different properties and, thus, different decomposition rates (k).

active fraction during the year at a rate that is difficult to predict as microorganisms die and are decomposed by other organisms. Weather also influences the amount of mineralization as freezing and thawing or drying and rewetting cycles cause a conversion of stable organic molecules, perhaps in some intimate association with clays, to highly labile molecules (Bartlett, 1981; Birch, 1958; Mack, 1963). In addition, there are different rates of mineralization for SOM present in particulate form, associated with various size aggregates, or directly associated silt or clay (Buyanovsky et al., 1994).

Another complicating factor is that SOM derived exclusively from plant residues is apparently more stable than that derived from manure (Christensen, 1992). This may help to explain the larger build-up and decreased activity of organic matter under cover-crop system than in manure amended soils found by Wander et al. (1994). In an extensive review of literature Christensen (1992) concluded that plant-derived SOM is more closely associated with a relatively stable silt fraction, while the partially decomposed materials and organisms present in manures become affiliated in a less stable relationship with clay.

Estimating Pesticide Adsorption Effects from Soil Organic Matter Tests

The degree of retention of a particular pesticide in a given soil is usually referred to as the distribution coefficient (K_d). Since SOM is one of the main soil constituents that adsorbs pesticides, it is common to compare pesticide retention by calculating the amount of pesticide adsorbed per unit of SOM ($K_{oc} = K_d(100)/\%$ organic – C). This value is used to estimate the adsorption of a pesticide by soils and modify recommended application rates of certain pesticides so that the higher the SOM, the greater is the application rate recommended. The following two assumptions, however, are made when using the concept of K_{oc} for this purpose: (i) organic matter is the overwhelming adsorber of pesticides in soil; and (ii) organic matter in all soils adsorbs a certain pesticide to the same extent. This means that K_{oc} values should be much the same for all soils; however, K_{oc} may vary greatly, even for soils with low amounts of clay, and tends to be positively correlated to SOM content (i.e., see Mallawatantri & Mulla, 1992; Seybold et al., 1994). Humus characteristics (original source of residues and degree of humification) may be responsible for these differences.

In one study, while the ranking of five pesticide K_{oc} estimates from the literature coincided with evaluations on a number of soils, approximately one-half of the pesticide–soil combinations were outside the mean ±1 standard deviation for values from the literature (Green et al., 1993). This indicates that while projecting relative environmental fates of pesticides from literature K_{oc} values may be useful, caution is needed when trying to estimate what might actually happen in a given soil.

Another significant limitation to using SOM to modify recommendation rates of pesticides, is the use of various SOM procedures by different laboratories. The weight-loss-on-ignition ashing procedure (WLOI) gives values significantly greater (up to 23%) than the Walkley-Black digestion (see Schulte & Hopkins, 1996, this publication). Although many laboratories using the WLOI procedure report calculated SOM based on regression equations with other methods, some report WLOI directly. The WLOI procedure also greatly overestimates the organic matter content of soils with significant amounts of free carbonates. Thus, uncorrected WLOI data may greatly overestimate organic matter levels in a given soil.

PROPOSED ALTERNATIVES TO ROUTINE SOIL ORGANIC MATTER TESTS

A number of alternative tests have been proposed that might be more accurate indicators of the contributions of SOM to N mineralization as well as to pesticide adsorption.

Estimating the Active Fraction

There has been considerable interest in developing methods to evaluate the active organic matter fraction. One of the general approaches has been to determine the amount and characteristics of *free* organic matter that is not intimately associated with mineral soil constituents. This is usually accomplished by dispersing soil and either (i) sieving out large particles (>50 μm) and then separating the sand from the organic material or (ii) subjecting the dispersed material to a solution with a specific density of 1.6 to 2.2 g cm^{-3} and recovering the material that floats (Christensen, 1992). The resulting organic matter from the two procedures are commonly referred to as particulate organic matter (POM) and light fraction (LF) organic matter respectively. The percentage of the soil's total C or N in these organic matter fractions can range from extremely low to close to 100%. In general, changes in POM and LF occur more rapidly than total SOM, giving an early indication of the probable direction of SOM change.

Compared with total C or total organic matter, the magnitude of active SOM fractions may be more susceptible to changes in agronomic practices; however, the greater variability of the active organic matter tests than the total SOM tests (Elliott et al., 1994) may not make them as useful as desired. At this time, there is no generally accepted alternative test to estimate SOM's fertility contributions for routine soil testing purposes.

Estimating the Well-Decomposed Fraction

There has been some interest in using an alternative soil test for estimating the well decomposed (humus) fraction in soil for purposes of modifying herbicide application rates. The proposed test consists of a NaOH extraction in place of either the Walkley–Black, H_2O_2 digestion, or WLOI procedures (Strek & Weber, 1983). The extracting solution consists of 0.2 M NaOH, 0.002 M DTPA, and 2% (v/v) ethanol and is followed by spectrophotometric analysis at 650 nm. This procedure gives higher SOM numbers than H_2O_2 digestion, but lower than WLOI (Strek & Weber, 1983). While there is good reason to expect that this test might be useful for estimating humified soil materials, there is no indication that it is actually better in predicting pesticide activity in soils than the other SOM tests. This procedure is currently being used in North Carolina (see Frank and Roeth, 1996, this publication).

Estimating Amounts of Biomass or Active Soil Organic Matter Fraction as a Guide to Soil Fertility or Quality

There is no generally accepted test for estimating soil quality or gauging changes in soil quality (Magdoff, 1995). There are, however, a number of tests involving the living fraction have been proposed for the evaluation of soil health or quality. These include soil biomass, enzyme activities, metabolic quotient (respiration per unit microbial biomass), population diversity, numbers of selected indicator organisms, and others (Linden et al., 1994; Dick, 1994; Turco et al., 1994; Sikora et al., 1996, this publication).

An example of the complications that can occur when evaluating the living fraction is that the total biomass may not be a good indicator of biological activity (Verhoef & Brussard, 1990). Because of the actions of grazers of microorganisms and predators of grazing organisms, populations of individual microorganisms may not increase greatly even when biological activity is high. In studies with return of different quantities and types of organic residues, long-term additions of high C/N ratio wheat straw increased microbial activities, but not populations and the addition of low C/N legume residues decreased microbial activity, but increased microbial populations (Gupta et al., 1994). With these conditions, higher ratios of POM-N/total-N was associated with lower POM C/N (Fig. 2–4). A positive correlation between microbial N and potentially mineralizable (or active) soil N (Duxbury & Nkambule, 1994) is an indication that conditions that promote the build-up of active organic matter also may promote increasing microbial biomass. But because the build-up of active organic matter is frequently at least partially caused by lower mineralization rates, this may also indicate that relatively high microbial populations can occur under conditions of reduced SOM mineralization.

Because of the complexities involved in interpreting changes in biological indicators, Sikora et al. (1996, this publication) suggest that changes in the quan-

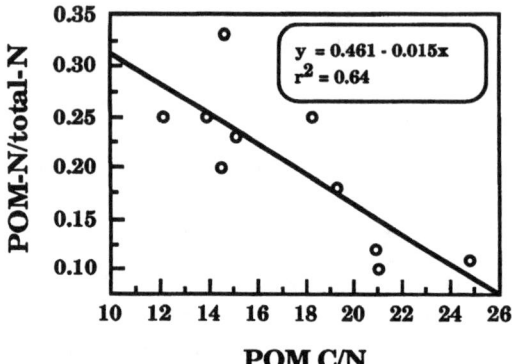

Fig. 2–4. The relationship between particulate organic matter (POM) C/N and the fraction total-N present as POM-N (from Gupta et al., 1994).

tity of particulate organic matter (POM, discussed above under the active fraction) may provide one of the best early indicators of changes in soil quality.

CONCLUSIONS

Caution is needed when interpreting currently used and proposed new SOM tests for making field recommendations. Although total SOM or certain SOM fractions influence numerous soil properties, the relationships may not be simple linear correlations. And while it seems beyond doubt that some part or parts of SOM are critical contributors to soil quality, it is still not clear which tests might provide useful indicators under the many varying and contrasting conditions found in the field.

REFERENCES

Birch, H.F. 1958. The effect of soil drying on humus decomposition and nitrogen availability. Plant Soil 10:9–31.

Bartlett, R.J. 1981. Oxidation-reduction status of aerobic soils. p. 77–102. *In* R.H. Dowdy et al. (ed.) Chemistry in the soil environment. ASA Spec. Publ. 40. ASA and SSSA, Madison, WI.

Buyanovsky, G.A., M. Aslam, and G.H. Wagner. 1994. Carbon turnover in soil physical fractions. Soil Sci. Soc. Am. J. 58:1167–1173.

Christensen, B.T. 1992. Physical fractionation of soil and organic matter in primary particle size and density separates. Adv. Agron. 20:1–90.

Dalal, R.C., and R.J. Mayer. 1986. Long-term trends in fertility of soils under continuous cultivation and cereal cropping in Southern Queensland: II. Total organic carbon and its rate of loss from the soil profile. Aust. J. Soil Res. 24:281–292.

Dick, R.P. 1994. Soil enzyme activities as indicators of soil quality. p. 107–124. *In* J.W. Doran et al. (ed.) Defining soil quality for a sustainable environment. SSSA Spec. Publ. 35. ASA and SSSA, Madison, WI.

Duxbury, J.M., and S.V. Nkambule. 1994. Assessment and significance of biologically active soil organic nitrogen. p. 125–146. *In* J.W. Doran et al. (ed.) Defining soil quality for a sustainable environment. SSSA Spec. Publ. 35. ASA and SSSA, Madison, WI.

Elliott, E.T., I.C. Burke, C.A. Monz, S.D. Frey, K.H. Paustian, H.F. Collins, E.A. Paul, C.V. Cole, R.L. Blevins, W.W. Frye, D.J. Lyon, A.D. Halvorson, D.R. Huggins, R.F. Turco, and M.V. Hickman. 1994. Terrestrial carbon pools: Preliminary data from Corn Belt and Great Plains Regions. p. 179–191. *In* J.W. Doran et al. (ed.) Defining soil quality for a sustainable environment. SSSA Spec. Publ. 35. ASA and SSSA, Madison, WI.

Frank, K.D., and F.W. Roeth. 1996. Using soil organic matter to help make fertilizer and pesticide recommendations. p. 33–40. *In* Soil organic matter: Analysis and interpretation. SSSA Spec. Publ. 46. SSSA, Madison, WI.

Green, R.E., R.C. Schneider, R.T. Gavenda, and C.J. Miles. 1993. Utility of sorption and degradation parameters from the literature for site-specific pesticide impact assessments. p. 209–225. *In* D.M. Linn et al. (ed.) Sorption and degradation of pesticides and organic chemicals in soil. SSSA Special Publ. 32. SSSA, Madison, WI.

Gupta, V.V.S.R., P.R. Grace, and M.M. Roper. 1994. Carbon and nitrogen mineralization as influenced by long-term soil and crop residue management systems in Australia. p. 193–200. *In* J.W. Doran et al. (ed.) Defining soil quality for a sustainable environment. SSSA Spec. Publ. 35. ASA and SSSA, Madison, WI.

Linden, D.R., P.F. Hendrix, D.C. Coleman, and P.C.J. van Vliet. 1994. Faunal indicators of soil quality. p. 91–106. *In* J.W. Doran et al. (ed.) Defining soil quality for a sustainable environment. SSSA Spec. Publ. 35. ASA and SSSA, Madison, WI.

Mack, A.R. 1963. Biological activity and mineralization of nitrogen in three soils as induced by freezing and drying. Can. J. Soil Sci. 43:316–324.

Magdoff, F.R. 1978. Influence of manure application rates and continuous corn on soil N. Agron. J. 70:629–632.

Magdoff, F.R. 1991. Field nitrogen dynamics: implications for assessing N availability. Commun. Soil Sci. Plant Anal. 22:1507–1517.

Magdoff, F.R. 1995. Defining soil quality for a sustainable environment. Book Rev. Am. J. Altern. Agric. 10:46.

Magdoff, F.R., and R.J. Bartlett. 1985. Soil pH buffering revisited. Soil Sci. Soc. Am. J. 49:145–148.

Mallawatantri, A.P., and D.J. Mulla. 1992. Herbicide adsorption and organic carbon contents on adjacent low-input versus conventional farms. J. Environ. Qual. 21:546–551.

Schnurer, J., M. Clarholm, and T. Rosswall. 1985. Microbial biomass and activity in an agricultural soil with different organic matter contents. Soil Biol. Biochm. 17:611–618.

Schulte, E.E., B.G. Hopkins. 1996. Estimation of soil organic matter by weight loss-on-ignition. p. 21–31. *In* Soil organic matter: Analysis and interpretation. SSSA Spec. Publ. 46. SSSA, Madison, WI.

Seybold, C.A., K McSweeney, and B. Lowery. 1994. Atrazine adsorption in sandy soils of Wisconsin. J. Environ. Qual. 23:1291–1297.

Sikora, L.J., C. Cambardella, V. Yakovchenko, and J. Doran. 1996. Assessing soil quality by testing organic matter. p. 41–50. *In* Soil organic matter: Analysis and interpretation. SSSA Spec. Publ. 46. SSSA, Madison, WI.

Strek, H.J., and J.B. Weber. 1983. Update on soil testing and herbicide recommendations. Proc. South. Weed Sci. Soc. 36:398–403.

Turco, R.F., A.C. Kennedy, and M.D. Jawson. 1994. Microbial indicators of soil quality. p. 73–90, *In* J.W. Doran et al. (ed.) Defining soil quality for a sustainable environment. SSSA Spec. Publ. 35. ASA and SSSA, Madison, WI.

Verhoef, H.A., and L. Brussard. 1990. Decomposition and nitrogen mineralization in natural and agroecosystems: The contribution of soil animals. Biogeochemistry 11:175–211.

Wander, M.M., S.J. Traina, B.R. Stinner, and S.E. Peters. 1994. Organic and conventional management effects on biologically active soil organic matter pools. Soil Sci. Soc. Am. J. 58:1130–1139.

Weber, J.B. 1994 Properties and behavior of pesticides in soil. p. 15–41. *In* R.C. Honeycutt and D.J. Schabacker (ed.) Mechanisms of pesticide movement to ground water. CRC Press, Boca Raton, FL.

Weber, J.B., and C.J. Peters. 1982. Adsorption, bioactivity, and evaluation of soil tests for alachlor, acetochlor, and metolachlor. Weed Sci. 30:14–20.

3 Estimation of Soil Organic Matter by Weight Loss-On-Ignition

E. E. Schulte

University of Wisconsin
Madison, Wisconsin

B. G. Hopkins

Kansas State University
Manhattan, Kansas

Chromic acid has been used widely in U.S. soil testing laboratories to measure oxidizable organic C as an estimate of soil organic matter (SOM). Concern for disposal of the Cr and hazards associated with its use, however, has prompted the search for alternative methods of estimating soil SOM.

A promising alternative is weight loss on ignition (WLOI). Estimation of SOM by WLOI is based upon measuring the weight loss from a dry soil sample due to high temperature ignition. Ideally, soil organic C would oxidize completely within a narrow temperature range at which weight loss from soil minerals is negligible. Unfortunately, this is not the case, so temperature selection is somewhat arbitrary but, at the same time, critical to minimize errors. The drying temperature should be high enough to remove a maximum amount of soil water but low enough to prevent loss of organic C. Analogously, the ignition temperature should be high enough to remove a maximum amount of organic C and low enough to minimize loss of other soil constituents. High temperature heating (>500°C) can result in errors arising from loss of CO_2 from carbonates, structural water from clay minerals, oxidation of Fe^{2+}, and decomposition of hydrated salts (Ball, 1964; Ben-Dor & Banin, 1989; Jackson, 1958). Heating at temperatures below 500°C, however, should eliminate many of these errors. Davies (1974), for example, compared the WLOI at 430°C for 17 British soils containing 9 to 36.5% $CaCO_3$ with SOM determined by the method of Walkley and Black (1934). The relationship (% SOM = −0.56 + 0.851 WLOI ; $r = 0.994**$) indicated no interference from the carbonates. Addition of $CaCO_3$ to a pH 6.8 soil to give soil/$CaCO_3$ ratios from 10:0 to 10:5 had no effect on WLOI.

Ben-Dor and Banin (1989) measured the WLOI at 400°C of 91 Israeli arid zone soil samples containing 0 to 74% $CaCO_3$. A regression equation relating WLOI to organic C by the Walkley and Black method gave: % WLOI = 0.383 + 2.049 (organic C); $r = 0.972**$. Thermogravimetric analysis of several natural

---

Copyright © 1996. Soil Science Society of America, 677 S. Segoe Rd., Madison, WI 53711, USA.
Soil Organic Matter: Analysis and Interpretation. SSSA Special Publication no. 46.

soils and two synthetic mixtures showed that heating the samples for 24 h at 105°C was necessary to remove hygroscopic water, which would otherwise be interpreted as SOM. The authors concluded that "quantitative organic matter determination in mineral soils of arid zones can be obtained by a simple WLOI method without pretreatment, except for drying at 105°C for 24 h, followed by ignition for 8 h at 400°C. None or very small errors are introduced by the presence of carbonates, smectite, kaolinite, and low amounts of iron oxides . . ."

The presence of gypsum ($CaSO_4 \cdot 2\ H_2O$) could present a problem with WLOI in soils of subhumid and arid regions (T. Jensen, 1993, personal communication). Gypsum contains 20.9% water. According to the Handbook of Chemistry and Physics (Lide, 1993), gypsum loses 1.5 H_2O at 128°C and the remaining H_2O at 163°C. Therefore, preheating at 150°C or higher should eliminate much of the problem with gypsum. Other hydrated salts also could introduce errors. Epsom salts ($MgSO_4 \cdot 7\ H_2O$) loses six H_2O molecules at 150°C and the remaining one at 200°C (Lide, 1993). Four H_2O molecules are lost from $CaCl_2 \cdot 6\ H_2O$ at 30°C, the remaining two at 200°C. Sodium salts, $Na_2CO_3 \cdot 10\ H_2O$, and $Na_2SO_4 \cdot 10\ H_2O$, both become anhydrous upon heating to 100°C and do not contribute to errors when using the WLOI procedure; however, $NaHCO_3$ decomposes to NaOH and CO_2 at 270°C (Lide, 1993). The effect of time of heating at 105 or 150°C on loss of water from these salts is not known.

Dehydroxylation of silicates begins at 350 to 370°C for many silicate clays; however, Na-montmorillonite, vermiculite, gibbsite, goethite, and brucite begin to lose crystal-lattice water at 150 to 250°C (Barshad, 1965). Water inside the cavities of O_2 surfaces is generally lost above 350°C; loss of water of hydration around exchangeable cations is completed at 150°C in Na-montmorillonite, Na-vermiculite, Ca- and Na-illite, gibbsite, and goethite. Temperatures of 250 to 350°C are needed to desorb layer water from Ca-montmorillonite, Ca-vermiculite, muscovite, phlogopite, biotite, pyrophyllite, kaolinite, halloysite, and antigorite. Consequently, some errors due to water loss by soil minerals between 105 and 360°C could be avoided by selecting a temperature above 105°C as the base temperature but below which organic matter decomposes. Time of heating both at the basal temperature and the ignition temperature must be controlled.

Schulte et al. (1991) studied the effects of heating time, sample size, and number of samples ignited at one time on WLOI at 360°C. The number of samples ignited at one time in a muffle furnace did not affect WLOI. For an organic soil containing 34% SOM, beaker size (20 or 50 mL) was unimportant, but WLOI increased as sample size decreased. Sample size was not significant for a mineral soil (3.6% SOM). Time of heating was significant for both soils. When the WLOI at 360°C for 2 h of 356 Wisconsin soils was compared with SOM determined by Walkley–Black titration (Schulte, 1988), the regression equation was: SOM = 1.04 WLOI − 0.36 ($R^2 = 0.97**$). Organic matter in these samples ranged from 0.1 to 54%. When only samples containing <10% SOM were included, the regression equation was: SOM = −0.33 + 0.973 ($R^2 = 0.90**$).

A number of papers have been published in the past decade on the use of WLOI to estimate soil organic matter (Table 3–1). As the slopes of the regression equations in Table 3–1 indicate, values for WLOI are higher than those for SOM. Hence, a regression equation is needed to estimate SOM from WLOI. The dif-

Table 3–1. Heating times and temperature reported in the literature for weight loss-on-ignition (WLOI).

Temperature	Time	y = bx + a†		R^2	n	Reference
		b	a			
°C	h					
360	2	0.73	−0.08	0.93**	63	Beverly et al., 1992
360	2	0.66	−0.04	0.84**	60	Beverly et al., 1992‡
360	2	1.04	−0.36	0.97**	356	Schulte et al., 1991
360	2	0.97	−0.33	0.90**	316	Schulte et al., 1991§
360	1	0.68	−0.50	0.90**	217	Storer, 1990, personal communication
375	16	0.79	−0.7	--	65	Ball, 1964
400	8	0.84	−0.32	0.97**	91	Ben-Dor & Banin, 1989
400	6	0.57	--	0.98**	55	Donkin, 1991
430	24	0.85	0.56	0.99**	17	Davies, 1974
450	12	0.90	−0.02	0.92**	174	David, 1988
450	12	¶	¶	0.89**	164	David, 1988#
450	16	0.78	−0.20	0.99**	38	Lowther et al., 1990
500	4	0.81	−1.47	0.98**	215	Storer, 1984
500	4	0.60	−0.33	0.87**	210	Storer, 1984§
600	6	0.70	−1.24	0.86**	60	Goldin, 1987††
600	6	0.72	−4.29	0.89**	12	Goldin, 1987#

** Significant at 0.01 probability level.
† y = % SOM; if organic C was reported, soil organic matter (SOM) was calculated assuming 58% C; x = % WLOI.
‡ % SOM <5% (the Gambia).
§ SOM <9.9%.
¶ % SOM = −4.72 + 1.40 % WLOI − 0.0443 (% WLOI)^2.
\# Forest floor litter.
†† Mineral soils (Canada).

ferences in slopes shown in Table 3–1 result from differences in heating times and temperature and, possibly, from differences in the nature of the clay and SOM fractions. Goldin (1987) and David (1988), for example, obtained different regression equations for mineral soils than for forest floor litter. Peters (1991, personal communication) and Storer (1984) obtained different slopes and intercepts when high organic matter samples were excluded from the regression analysis.

Soon and Abboud (1991) compared six methods of estimating organic C in 28 A_p and 11 subsoil (>15 cm) samples of agricultural soils in northwest Alberta. These methods included: (i) Walkley–Black (1934); (ii) Tinsley method (digest at 165°C in 5 mL 0.17 M K_2CrO_7 + 10 mL of 5:1 (v/v) H_2SO_4 - H_3PO_4 for 30 min); (iii) Mebius method (digest at 155°C); (iv) spectrophotometric analysis after digestion at 135°C for 30 min, using sucrose standards; (v) WLOI of air-dry samples for 2 h in an oven preheated to 250°C, followed by 16 h heating at 375°C; and (vi) dry combustion using a C analyzer. There was close correlation among all six procedures ($r > 0.95$**) in spite of the fact that air-dry samples were used for WLOI.

The purpose of the present study was to compare basal heating temperatures of 105 and 150°C in the WLOI procedure and to investigate the rate of water loss from gypsum at these temperatures.

MATERIALS AND METHODS

Soils

Several groups of plow-layer soil samples were included in a data set of 177 samples: 50 from the University of Wisconsin-Extension Soil and Plant Analysis Laboratory, Madison; 50 from the University of Wisconsin-Extension Soil and Forage Analysis Laboratory, Marshfield; 10 collected for laboratory course work; 37 from Wisconsin forest nurseries; 13 other forest soil samples, including 3 from Iceland; and 17 from Kansas. All samples were air dried and ground to pass a 14-mesh sieve (1.30-mm openings).

Organic Matter Analysis

Soil organic matter (SOM) was determined by the method of Walkley and Black (1934), as described by Schulte (1988). Organic matter was calculated assuming 58% C in SOM and that 77% of the organic C was oxidized.

Loss-On-Ignition

A 10-g soil sample was placed in a tared 50-mL Pyrex beaker and dried for 2 h at 105°C in a Thermolyne Model FA1740 furnace with an internal chamber measuring 24 cm wide, 22 cm high, and 34 cm deep. The temperature was controlled with a Furnatrol regulator. Because the regulator cycled current to the furnace intermittently until the set temperature was reached, residual heat in the ceramic heating elements raised the oven temperature above the set point by several degrees. Therefore, the furnace was preheated by setting the controller to a lower temperature and then gradually increasing the setting until the desired temperature was obtained. At this point, the controller maintained the desired temperature within ±5°C. After 2 h at 105°C, the samples were weighed while hot to ±0.001 g. The furnace temperature was then raised to 150°C, at which temperature the samples were heated for 2 h, and again weighed while hot. Finally, the oven temperature was elevated to 360°C for 2 h. The oven was then shut off and allowed to cool overnight to 100 to 150°C before weighing the samples. Loss-on-ignition was calculated using both the 105 and 150°C weights as the base weight.

The effect of heating temperature on WLOI was studied by heating three soil samples and a sample of calcareous loess in triplicate for 2 h at 200, 300, 400, 500, 600, and 700°C in porcelain crucibles. The samples were weighed to ±0.001 g before being returned to the furnace at the next higher temperature. Weight loss was calculated using the oven-dry (105°C) weight as the base. The pH of the samples was measured with a glass electrode pH meter in a 2:1 water/soil suspension after equilibrating for 30 min.

Loss of Water from Gypsum

Water loss from gypsum and a Fayette silt loam (fine-silty, mixed, mesic Typic Hapludalf) was measured in triplicate in 5-g samples of soil or C.P. grade gypsum heated in 50-mL beakers alone or as equal mixtures by weight at 105 and 150°C in a forced-air drying oven over a period of 25 h (105°C) or 6 h (150°C). Water content was calculated from weight loss at unequal intervals over these periods using air-dry weight as the base. The Fayette soil contained 1.76 ±0.26% SOM determined by the Walkley–Black (1934) method. It was used because it was available as a teaching laboratory standard of known SOM.

RESULTS AND DISCUSSION

Influence of Basal Temperature on Weight Loss-On-Ignition at 360°C

When all 177 soil samples were taken together, simple regression analysis showed no improvement in WLOI by heating at 150°C instead of 105°C to remove residual moisture. The regressions equations for the two temperatures and mean values of SOM and WLOI were:

	R^2	n	OM	$WLOI_{105}$	$WLOI_{150}$
			\-\-\-\-\-\-\- % \-\-\-\-\-\-\-		
OM = 0.23 + 0.773 $WLOI_{105}$	0.935**	177	3.53	4.27	
OM = 0.49 + 0.796 $WLOI_{150}$	0.936**				3.81

The intercepts and slopes of the two equations varied because of differences in weight loss at the two temperatures, as expected, but the coefficients of determination (R^2) are nearly identical. Removal of six outliers, four containing >10% SOM and two forest soils from Iceland, did not improve the coefficients of determination:

	R^2	n	OM	$WLOI_{105}$	$WLOI_{150}$
			\-\-\-\-\-\-\- % \-\-\-\-\-\-\-		
OM = −0.77 + 0.845 $WLOI_{105}$	0.933**	171	3.10	3.74	
OM = 0.11 + 0.905 $WLOI_{150}$	0.943**				3.30

The regressions equations were improved somewhat when forest soils and soils from Kansas were separated from the Wisconsin agricultural soils (Fig. 3–1, 3–2, 3–3; note that SOM is plotted on the *x*-axis in Fig. 3–1, 3–2, and 3–3 so that the data for each temperature can be plotted on the same graph. The regression equations, however, have SOM as the dependent variable.)

The best improvement in the coefficient of determination when using 150°C as the base for WLOI calculations was obtained with the soils from Kansas (Fig. 3–3). Although this data set has only 17 samples, the increase in R^2 from 0.958** to 0.982** is large enough to encourage additional research on a larger number of samples. A possible explanation for failure to obtain better improvement in R^2 at 150°C is discussed below.

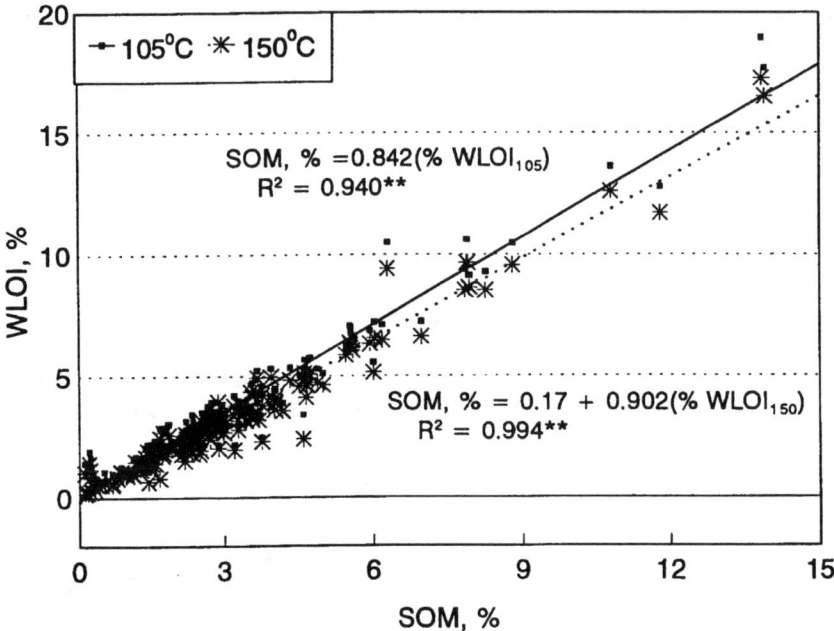

Fig. 3–1. Weight loss on ignition (WLOI) in Wisconsin agricultural soils.

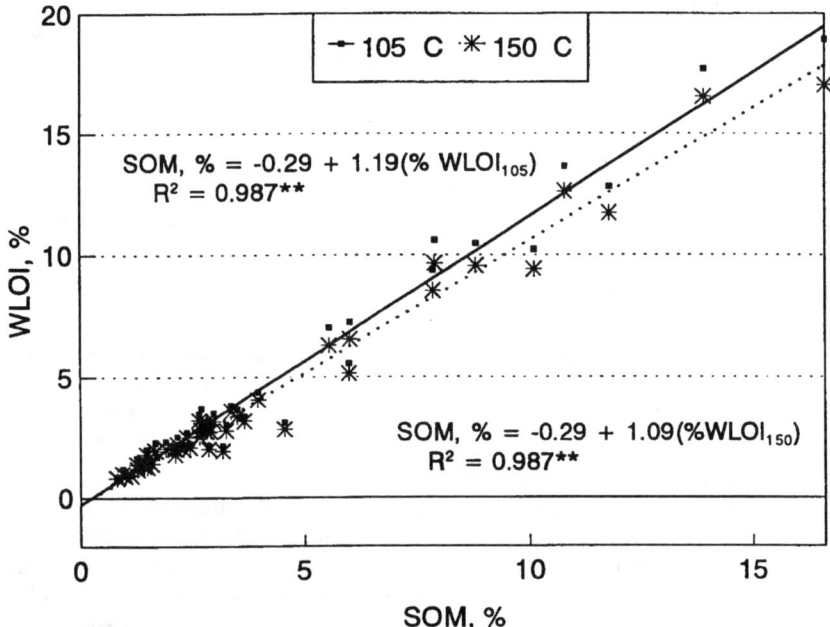

Fig. 3–2. Weight loss on ignition (WLOI) in forest soils.

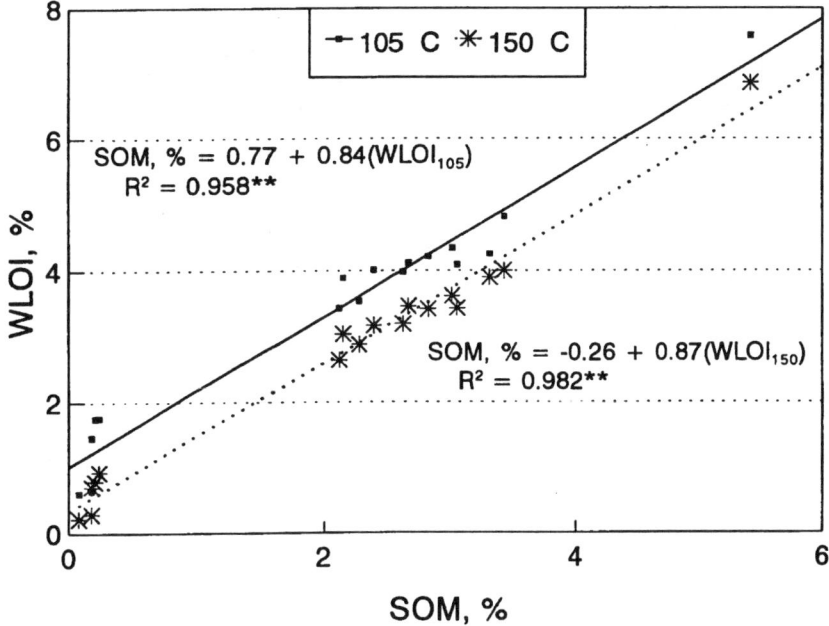

Fig. 3–3. Weight loss on ignition (WLOI) in 17 Kansas soil samples.

Loss of Moisture from Gypsum at 105 and 150°C

Water loss from gypsum in gypsiferous soils would give erroneously high results for SOM. The range in temperature and ease with which gypsum loses water was unknown. Also, some water of hydration may be lost from soil clays between 105 and 150°C, as discussed above. Therefore, the rate of water loss from gypsum was studied at 105 and 150°C. Gypsum contains 20.9% water. One-half of the water was lost in 2 h at 105°C and all at 150°C (Fig. 3–4). More than 80% of the water was lost in 6 h at 105°C, and complete dehydration occurred in 25 h. These data explain why there was not an even greater difference in WLOI calculated from 105 and 150°C bases. Mixing the gypsum with the soil did not affect the rate of water loss (Fig. 3–4). To ensure complete dehydration of gypsiferous soils, a minimum of 2 h at 150°C or overnight at 105°C is necessary. For other soils, the more standard 105°C base should suffice. For soils containing clays that might not dehydrate completely in 2 h at 105°C, a study of time and temperature effects on moisture loss would help in the selection of optimum drying conditions.

Effect of Ignition Temperature on Weight Loss-On-Ignition

Because SOM oxidizes across a fairly wide range in temperature, the optimum temperature for ignition is somewhat arbitrary. The objective is to find a temperature at which oxidation is nearly complete but weight loss from soil minerals is negligible. Weight loss from three soil samples and a sample of calcare-

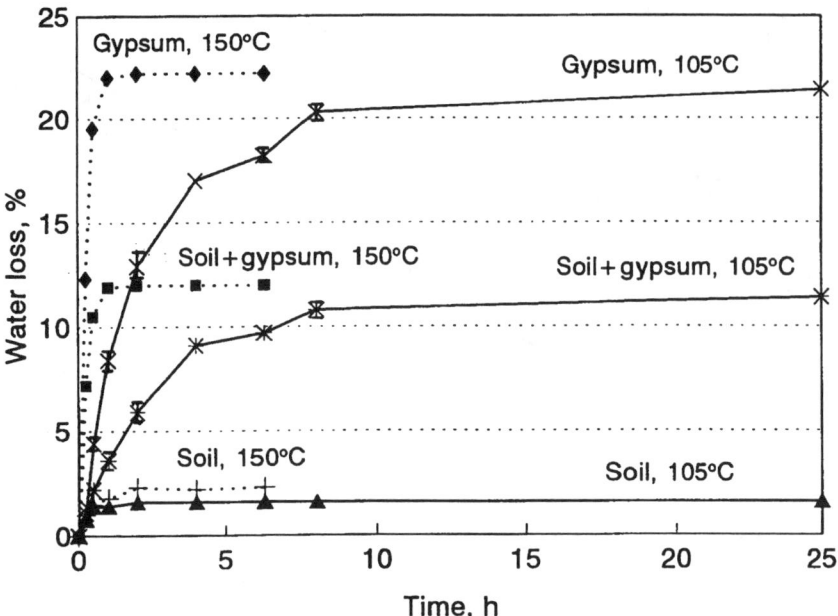

Fig. 3–4. Water loss from gypsum, Fayette silt loam, and mixtures (1:1 w/w) of gypsum and soil as a function of time and temperature.

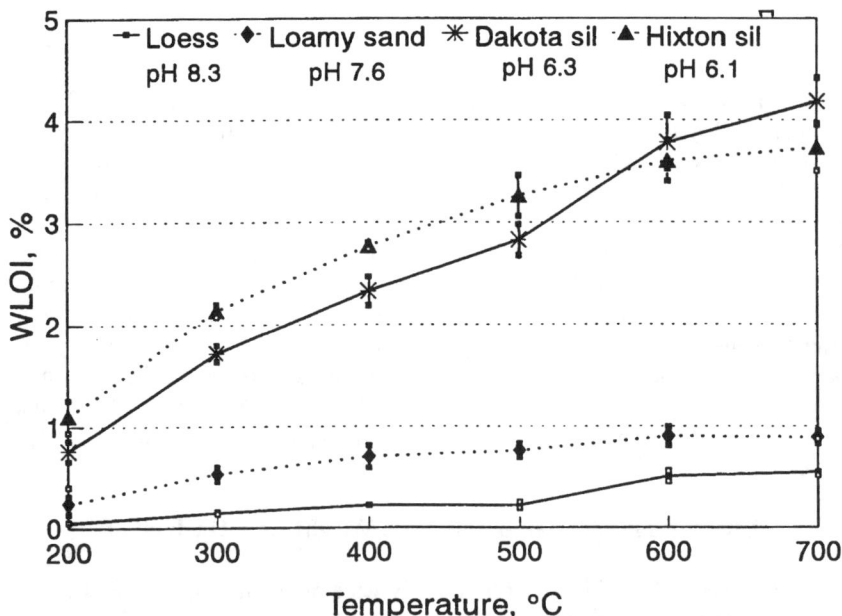

Fig. 3–5. Influence of heating temperature on weight loss on ignition (WLOI; based on 105°C weight).

ous loess was studied between 200 and 700°C. Weight loss from each sample was curvilinear up to 500°C (Fig. 3–5). Above this temperature there was an abrupt increase in WLOI from the loess and the Dakota sandy loam (fine-loamy over sandy or sandy-skeletal, mixed, mesic Typic Argiudoll) as carbonates in these calcareous samples decomposed. Thus, any temperature between 250 and 500°C could be used to measure WLOI, but the temperature selected must be controlled. Higher temperatures give greater weight loss and, therefore, are more sensitive because WLOI is calculated as the difference in weight at 105°C and at the temperature chosen. A temperature of 360°C as used by Storer (1990, personal communication) has become accepted by the North central region (Schulte, 1988). Donkin (1991) compared WLOI of 15-g samples of 55 noncalcareous South African soils ignited for 6 h between 100 and 900°C. The best comparison with organic C (Walkley–Black) was obtained at 450 ($r = 0.966**$). Although results at other temperatures between 250 and 900°C gave almost equally good correlations ($r = 0.958**$ to $0.971**$), the 450 temperature gave the lowest standard error of prediction.

Effect of Heating Time on Weight Loss-On-Ignition

Schulte et al. (1991) compared WLOI in a mineral and an organic soil when heated at 360°C for 2, 4, and 16 h. Significant differences in WLOI were found for each time period in both soils. Ben-Dor and Banin (1989) showed by differential thermal and thermogravimetric analyses that dehydration of phyllosilicates and gypsum takes place between 100 and 200°C and between 200 and 300°C for polygorskite and halloysite. For research purposes, they recommended drying at 105°C for 24 h, followed by ignition at 400°C for 8 h.

For routine analysis of large numbers of samples, accuracy in the analysis of SOM to the second decimal place is unnecessary. The results of SOM analyses from laboratories in the USA are used primarily to estimate N availability, adjust herbicide rates, and to estimate reserve acidity and buffering. For these purposes, a more rapid estimate of SOM is needed. The research reported in this chapter was conducted with a furnace. Routine soil testing labs typically use high-temperature forced-air ovens. The steady supply of O_2 might shorten heating time, but whatever time is selected needs to be controlled. By using shorter heating times and correlating WLOI with SOM determined by conventional methods, a reliable SOM test should be possible.

Correlation of Weight Loss-On-Ignition with Soil Organic Matter

Organic matter is estimated from WLOI by use of a regression equation relating WLOI to some method of measuring SOM. Soil testing laboratories typically measure SOM by the method of Walkley and Black (1934), assuming that the procedure measures easily oxidizable organic C. A correction factor is used to convert this value to total organic C, and another factor converts organic C to organic matter. In the SOM method as originally proposed, Walkley and Black (1934) recovered an average of 76% of the organic C in 20 soil samples, but recovery values ranged from 60 to 86%. Kaufmann (1991, unpublished data) recovered 49 to 96% (average = 78%) of the organic C in 96 Wisconsin soils by this method compared with the procedure of Nelson and Sommers (1975). Walkley (1947) discusses sources of variation inherent in the procedure. These include differences due to digestion conditions, certain inorganic soil constituents (especially Cl, MnOx, and Fe), and the composition of soil SOM itself. Because of these inherent sources of error, it would be best to correlate WLOI with organic C determined by means of a C analyzer, taking precautions to remove interferences from inorganic C, and eliminate the use of correction factors.

CONCLUSIONS

Organic matter in soil can be estimated with reasonable accuracy by WLOI. Samples must be dried to remove moisture. Drying for 24 h at 105°C removed all of the water from gypsum, but 50% was lost in 2 h. Gypsum was completely dehydrated in 2 h at 150°C. The latter basal temperature might be useful also for soils containing hydrated clays. Careful attention must be given to temperature and duration of heating for WLOI analysis. Results must be calibrated against organic C, preferably determined with a C analyzer.

For research purposes, samples are normally cooled in a desiccator before weighing. This is not practical for routine analysis of large numbers of samples. Nevertheless, it is essential that samples are not allowed to rehydrate after ignition. For that reason, we recommend weighing directly from the oven or furnace after the temperature has dropped below 150°C, but before it reaches 105°C.

ACKNOWLEDGMENTS

This research was supported in part by a grant from the Wisconsin Fertilizer Research Council, which is gratefully acknowledged. We also extend our thanks to Dr. T. Jensen, Department of Agriculture, State of Nebraska, for valuable suggestions concerning potential sources of error in the WLOI procedure. Thanks also are due to Eva M. Schulte, Department of Soil Science, University of Wisconsin, Madison, who assisted with the Walkley–Black organic matter analysis.

REFERENCES

Ball, D.F. 1964. Loss-on-ignition as an estimate of organic matter and organic carbon in noncalcareous soils. J. Soil Sci. 15:84–92.

Barshad, I. 1965. Thermal analysis techniques for mineral identification and mineralogical composition. p. 699–742. In C.A. Black et al. (ed.) Methods of soil analysis. Part I. Agron. Monogr. 9. ASA, CSSA, and SSSA, Madison, WI.

Ben-Dor, E., and A. Banin. 1989. Determination of organic matter content in arid-zone soils using a simple "loss-on-ignition" method. Commun. Soil Sci. Plant Anal. 20:1675–1695.

Beverly, R.B., J.B. Peters, S. Njie, and E.E. Schulte. 1992. Application of soil testing in The Gambia. Commun. Soil Sci. Plant Anal. 23:2339–2346.

David, M.B. 1988. Use of loss-on-ignition to assess soil organic carbon in forest soils. Commun. Soil Sci. Plant Anal. 19:1593–1599.

Davies, B.E. 1974. Loss-on-ignition as an estimate of soil organic matter. Soil Sci. Soc. Am. Proc. 38:150–151.

Donkin, M.J. 1991. Loss-on-ignition as an estimate of soil organic matter in A-horizon forestry soils. Commun. Soil Sci. Plant Anal. 22:233–241.

Goldin, A. 1987. Reassessing the use of loss-on-ignition for estimating organic matter content in non-calcareous soils. Commun. Soil Sci. Plant Anal. 18:1111–1116.

Jackson, M.L. 1958. Soil chemical analysis. Prentice-Hall, Englewood Cliffs, NJ.

Lide, D.R. (ed.). 1993. Handbook of chemistry and physics. CRC Press, Ann Arbor, MI.

Lowther, J.R., P.J. Smethurst, J.C. Carllyle, and E.K.S. Nambiar. 1990. Methods for determining organic carbon on podzolic sands. Commun. Soil Sci. Plant Anal. 21:457–470.

Nelson, D.W., and L.E. Sommers. 1975. A rapid and accurate procedure for estimation of organic carbon in soils. Proc. Indiana Acad. Sci. 84:456–462.

Schulte, E.E. 1988. Recommended soil organic matter tests. p. 29–31. *In* W.C. Dahnke (ed.) Recommended chemical soil test procedures for the North Central Region. NCR Publ. no. 221 (Revised). Coop. Ext. Serv., North Dakota State Univ., Fargo.

Schulte, E.E., C. Kaufmann, and J.B. Peters. 1991. The influence of sample size and heating time on soil weight loss-on-ignition. Commun. Soil Sci. Plant Anal. 22:159–168.

Soon, Y.K., and S. Abboud. 1991. A comparison of some methods for soil organic carbon determination. Commun. Soil Sci. Plant Anal. 22:943–954.

Storer, D.A. 1984. A simple high sample volume ashing procedure for determination of soil organic matter. Commun. Soil Sci. Plant Anal. 15:759–772.

Walkley, A. 1947. A critical examination of a rapid method for determining organic carbon in soils - effect of variations in digestion conditions and of inorganic soil constituents. Soil Sci. 63:251–264.

Walkley, A., and I.A. Black. 1934. An examination of the Degtjareff method for determining soil organic matter, and a proposed modification of the chromic acid titration method. Soil Sci. 37:29–38.

4 Using Soil Organic Matter to Help Make Fertilizer and Pesticide Recommendations

K. D. Frank

University of Nebraska
Lincoln, Nebraska

F. W. Roeth

University of Nebraska
Clay Center, Nebraska

Soil organic matter (SOM) influences the physical and chemical properties of soils. Soil structure, water holding capacity, infiltration of water and air, reducing soil erosion, and influencing pesticide efficacy are all influenced by SOM (Gregorich et al., 1993). From a plant nutrient standpoint, SOM serves as a store house for N, P, S, and Zn (Sikora & Stott, 1996, unpublished data in press; Mehring & Bennett, 1950; Thompson et al., 1954).

Research has shown that SOM varies with location; consequently, the quantity of inorganic N derived from SOM through mineralization and subsequent uptake by plants will vary. On average, N uptake by plants from mineralization of SOM is estimated to be 150 g kg^{-1} (15%) of total plant N requirement (National Academy of Sciences, 1987). Thus, to gain information on the use of SOM content in determining nutrient and pesticide recommendations for crop production, a survey was sent to commercial and university soil testing laboratories across the USA. Results of this survey are reported here.

SOIL ORGANIC MATTER AND NUTRIENT RECOMMENDATIONS

Survey responses were received from 25 universities, one state department of agriculture, and nine commercial laboratories. Laboratory locations and information on whether SOM is run routinely, as an option, or not at all, is shown in Table 4–1. Of the university laboratories surveyed, two do not determine SOM.

The University of Arkansas laboratory does not determine SOM because SOM levels are generally low throughout the state (Dr. Wayne Sabbe, Dep. of Agronomy, Univ. of Arkansas, Fayetteville, April 1994, personal communication).

Copyright © 1996. Soil Science Society of America, 677 S. Segoe Rd., Madison, WI 53711, USA. *Soil Organic Matter: Analysis and Interpretation.* SSSA Special Publication no. 46.

Table 4–1. Location of laboratories surveyed relative to determination and use of soil organic matter (SOM) in fertilizer and pesticide recommendations.

Location†	Soil organic matter determined	
	Routinely	Option
University of Arkansas‡		
University of California College of Agriculture and Environmental Sciences, Davis		Yes
Colorado State University College of Agricultural Sciences	Yes	
Clemson University		Yes
University of Florida		Yes
University of Georgia		Yes
Iowa State University of Science and Technology	Yes	
Kansas State University		Yes
University of Kentucky		Yes
Louisiana State University		Yes
Michigan State University		Yes
University of Minnesota		Yes
University of Missouri	Yes	
University of Nebraska	Yes	
North Carolina†	Yes	
Ohio State University		Yes
Oklahoma State University		Yes
Oregon State University		Yes
Pennsylvania State University		Yes
Purdue University		(uses Ohio State)
South Dakota State University		Yes
University of Tennessee		Yes
Texas A&M University		Yes
Utah State University of Agriculture and Applied Science		Yes
Virginia Polytechnic Institute and State University		Yes
University of Wisconsin	Yes	
Commercial Laboratories		
California		Yes
Oregon	Yes	
Idaho	Yes	
Illinois	Yes	
Nebraska (4 Laboratories)	Yes	
Wisconsin	Yes	

† All university laboratories are located in the Department of Agronomy.
‡ Arkansas does not run SOM.
§ North Carolina's Laboratory is located in the State Dep. of Agriculture.

The North Carolina State Department of Agriculture Agronomic Division Soil Testing Laboratory analyzes soils for humic matter (Tucker & Messick, 1995). Humic matter represents the portion of humic and fulvic acids that are chemically reactive within soils. Humic matter is used to classify soils into three major categories: mineral, 0 to 35 g kg^{-1} (3.5%); mineral-organic, 36 to 52.5 g kg^{-1} (3.6 to 5.25%); and organic, 52.6 g kg^{-1} (5.26%) or greater (Dr. M. Ray Tucker, North Carolina State Dep. of Agriculture, Raleigh, April 1994, personal communication). These categories are used for adjusting recommendations for rates of lime, P, and Cu (Tucker & Messick, 1995). The target pH factored into lime recommendations by soil category is: mineral, 6.0 to 6.5; mineral-organic,

Table 4–2. Parameters in addition to soil organic matter (SOM) used in making nutrient recommendations for various crops by university laboratories that routinely analyze for SOM.

Laboratory	Crop	Nutrient	Parameters in addition to SOM
Colorado State University College of Agricultural Sciences	Corn	N	Expected yield
University of Nebraska	Corn	N	Expected yield
University of Nebraska	Sorghum	N	Expected yield
University of Nebraska	Alfalfa	S	Soil texture, SO_4–S in irrigation water
University of Nebraska	Corn	S	Same as for alfalfa
University of Missouri	Corn	N	Soil texture and cation-exchange capacity
University of Wisconsin	Corn	N	Growing degree days and availability of irrigation

5.5; and 5.0 for organic soils. Phosphorus, Zn, and Cu recommendations use the soil categories based on $10 \times HM$ g kg^{-1} (HM%) in combination with an availability index (Tucker & Messick, 1995).

Laboratories that analyze for SOM use the value in different ways. Some of these ways are: soil categories for making lime recommendations, N recommendations for various crops, and S recommendations.

Soil Categories

The Ohio State and Purdue Universities use SOM content to separate soils into two categories for making lime recommendations: mineral, 0 to 200 g kg^{-1} (0 to 20%); and muck, >200 g kg^{-1} (20%; M.E. Watson, Ohio State Research and Extension Analytical Laboratory, Wooster, April 1994, personal communication). Lime recommendations are based on a water pH and SMP buffer pH for the muck and mineral soils respectively (Jay Johnson, Dep. of Agronomy, Ohio State Univ., Columbus, April 1994, personal communication).

Fertilizer Nitrogen Recommendation

SOM content is determined routinely by university soil testing laboratories in Nebraska, Colorado, Missouri, and Wisconsin. Some examples of other parameters used by these laboratories, in conjunction with SOM, for various nutrient recommendations are shown in Table 4–2. The amount of credit given for SOM and other parameters varies by state. The use of SOM content for fertilizer N recommendations will be discussed by individual states.

Nebraska

Nebraska has groundwater control areas designated by the Nebraska Department of Environmental Quality based on the NO_3–N) content of the groundwater. Crop consultants, commercial soil testing laboratories, or anyone else making fertilizer N recommendations for crops in groundwater control areas are obligated to use University of Nebraska fertilizer N recommendations.

Nebraska uses SOM content as a variable, as shown in Eq. [1], to develop fertilizer N recommendations for corn (*Zea mays* L.; Penas et al., 1994).

$$N_{rec} = 35 + (1.2 \times EY) - (8 \times NO_3-N)$$
$$- (0.14 \times EY \times SOM) - \text{other N credits} \qquad [1]$$

N_{rec} = fertilizer N recommendation in lbs N/acre
EY = expected yield, bu/ac
NO_3-N = average NO_3-N in root zone to depth of 18 inches or greater
SOM = % organic matter of the top soil
Other credits = legume, manure, or NO_3-N in irrigation water

When EY and N_{rec} are in kilograms per hectare and SOM is in grams per kilogram, Eq. [1] becomes Eq. [1a].

$$N_{rec} = 39.2 + (0.02141 \times EY)$$
$$- (8.96 \times NO_3-N) - (0.00025 \times EY \times SOM) - \text{Other N credits} \qquad [1a]$$

N_{rec} = fertilizer N recommendation in kg ha^{-1}
EY = expected yield, kg ha^{-1}
NO_3-N = average NO_3-N in root zone to depth of 46 cm or greater in mg kg^{-1} (ppm)
SOM = organic matter content in top soil in g kg^{-1}
Other credits = legume, manure, or NO_3-N in irrigation water

The expected range of adjustment in Nebraska fertilizer N recommendations due to SOM can be estimated from the SOM term in Eq. [1] and [1a]. Reasonable SOM ranges are 8 to 30 g kg^{-1} (0.8 to 3%) and EY values of 6271 to 12542 kg ha^{-1} (100 to 200 bu/acre) respectively. For EY = 6271 kg ha^{-1} (100 bu/acre), the N adjustment is 12 and 47 kg ha^{-1} (11 and 42 lb N/acre) when SOM = 8 and 30 g kg^{-1} (0.8 and 3%), respectively. For EY = 12 542 kg ha^{-1} (200 bu/acre) the N adjustment varies from 25 to 94 kg ha^{-1} (22 to 84 lb N/acre) for SOM values of 8 to 30 kg^{-1} (0.8 to 3%). The Colorado State University Department of Agronomy soil testing laboratory uses the University of Nebraska fertilizer N recommendation for corn (G.A. Peterson, Colorado State Univ., Fort Collins, April 1994, personal communication).

Nebraska also uses SOM to adjust fertilizer N recommendations for grain sorghum [*Sorghum bicolor* [L.] Moench] as shown in Eq. [2] (Sander & Frank, 1980).

$$N_{rec} = 50 + (1.1 \times EY) - (20 \times [\% \text{ SOM} - 1]) - \text{other N credits} \qquad [2]$$

N_{rec} = fertilizer N recommendation in lb N/acre
EY = expected yield, bu/ac
SOM = % organic matter of the top soil
Other credits = legume, manure, NO_3-N in irrigation water and NO_3-N in root zone estimated to a depth of 6 feet in lb N/acre

The fertilizer N adjustment for SOM is positive when %SOM is <1 as calculated by the factor 20 × (%SOM − 1).

When EY and N_{rec} are in kilogram per hectare and SOM is in gram per kilogram, Eq. [2] becomes Eq. [2a].

$$N_{rec} = 56 + (0.0196 \times EY) - (2.24 [SOM- 10]) - \text{Other N credits} \quad [2a]$$

N_{rec} = fertilizer N recommendation in kg ha^{-1}
EY = expected yield, kg ha^{-1}
SOM = organic matter content of top soil in g kg^{-1}
Other credits = legume, manure, NO_3–N in irrigation water and NO_3–N in root zone estimated to a depth of 182 cm in mg kg^{-1}

On sandy textured soils in Nebraska, SOM is used to adjust S recommendations for corn and alfalfa (*Medicago sativa* L.). The adjustment is based on three levels of soil test sulfate–S, [<6, 6 to 7.9, and 8 or greater mg kg^{-1} (ppm) respectively]; two SOM levels, 10 or less and more than 10 g kg^{-1} (1% or less and more than 1%); and two levels of S in irrigation water, 6 mg L^{-1} (ppm) or less and more than 6 mg L^{-1} (ppm; Frank & Knudsen 1989).

Missouri

Commercial soil testing laboratories in Missouri associated with the Agriculture Stabilization Conservation Service (ASCS) cost share programs are required to use University of Missouri recommendations for fertilizer N as well as other nutrients (James Brown, University of Missouri, Columbia, April 1994, personal communication). The University of Missouri soil testing laboratory uses SOM, soil texture, and cation-exchange capacity (CEC) to adjust the fertilizer N rate in increments for grain crops, sugarbeet (*Beta vulgaris* L.), grass pastures, and other commercial crops (Buchholz, 1983). The fertilizer N adjustment for corn ranges from 22.4 kg ha^{-1} (20 lbs N/acre) on a sandy soil, CEC 10 or less and SOM 5 g kg^{-1} (0.5%), or less, to 90 kg ha^{-1} (80 lbs N/acre) on a silt loam, CEC of 10 to 18, and SOM content >40 g kg^{-1} (4%; Buchholz, 1983).

Wisconsin

Commercial soil testing laboratories in Wisconsin associated with the ASCS cost share program must use University of Wisconsin analysis methods and fertilizer recommendations. In addition, the commercial laboratories must participate in a soil sample check program managed by the University of Wisconsin soil testing laboratory (Sherry Combs, Univ. of Wisconsin, Madison, April 1994, personal communication).

The University of Wisconsin soil testing laboratory uses SOM (in increments) in combination with soil yield potential to arrive at the fertilizer N recommendation for corn (Kelling et al., 1991). Soil yield potential is a function of soil texture, growing degree days (GDD), and irrigation availability. The adjustment in the fertilizer N recommendation for corn when SOM content changes from <20 to >100 g kg^{-1} (<2 to >10%) ranges from 45 kg ha^{-1} (40 lb N/acre) on a nonirrigated sandy soil to 112 kg ha^{-1} (100 lb N/acre) on a nonsandy soil with

GDD >2300. They also use SOM in determining lime requirement as well as fertilizer N for other grain, oil, and vegetable crops (Kelling et al., 1991).

Minnesota

The University of Minnesota's Department of Agronomy uses either residual NO_3–N in the 0- to 60-cm (0- to 24-inch) depth or SOM to make fertilizer N recommendations for grain crops (Rehm et al., 1993). The preferred method is to use residual NO_3–N in combination with yield goal and N credit for previous legume crops as shown in Eq. [3].

$$N_{rec} = 1.2 \times YG - STN_{(0-24in)} - N_{PC} \quad [3]$$

N_{rec} = amount of fertilizer N needed, lb/acre
YG = realistic yield goal, bu/acre
$STN_{(0-24\ in)}$ = amount of NO_3–N measured in the 0- to 24-in depth by using the soil nitrate test, lb/acre
N_{PC} = N credits, if any, for previous crops in the rotation, lb/acre

When EY and N_{rec} are in kilogram per hectare, Eq. [3] becomes Eq. [3a].

$$N_{rec} = 0.02141 \times YG - STN_{(0-60\ cm)} - N_{PC} \quad [3a]$$

N_{rec} = amount of fertilizer N needed, kg ha^{-1}
YG = realistic yield goal, kg ha^{-1}
$STN_{(0-60\ cm)}$ = amount of NO_3–N measured in the 0- to 60-cm depth by using the soil nitrate test, kg ha^{-1}
N_{PC} = N credits, if any, for previous crops in the rotation, kg ha^{-1}

When residual NO_3–N is not available, the fertilizer N recommendation is based on a realistic yield goal, previous legume crop, and SOM level. When SOM is used, the soils are placed into categories, i.e., low, <30 g kg^{-1} (<3%), and medium and high SOM, 30 g kg^{-1} or more (3%). Table 4–3. shows an example of the N_{rec} for corn. The use of SOM in making fertilizer N recommendations for grain, vegetable, forage, and oil crops is given in greater detail elsewhere (Rhem et al., 1993).

Table 4–3. Example of University of Minnesota N recommendations for corn as a function of yield goal, soil organic matter (SOM) and previous crop. After Rehm et al., 1993.

Previous crop	SOM level	Yield goal (bu/acre)				
		91–110	111–130	131–150	151–170	171–190
		N to apply kg ha^{-1} (lb/acre)				
Alfalfa	<30 g kg^{-1} (3%)	0	0	0	33.6 (30)	56.0 (50)
	30 g kg^{-1} (3%) or more	0	0	0	0	22.4 (20)
Soybean	< 30 g kg^{-1} (3%)	56.0 (50)	89.6 (80)	123.2 (110)	156.8 (140)	179.2 (160)
	30 g kg^{-1} (3%) or more	33.6 (30)	67.2 (60)	89.6 (80)	123.2 (110)	145.6 (130)

SOIL ORGANIC MATTER AND PESTICIDE RECOMMENDATIONS

Soil organic matter is a major factor affecting the recommended application rate of soil-applied pesticides, particularly herbicides. Usually, herbicide rates are adjusted for soil differences, whereas insecticides are not. Pesticide sorption by soil colloids lowers the amount dissolved in the soil-water phase. The relative partitioning of a pesticide between the sorbed and dissolved phases (adsorption coefficient) affects its availability for plant uptake, leaching, and runoff. Pesticides with high adsorption coefficients are normally not economical for direct soil application. For example, the herbicide glyphosate has a high adsorption coefficient so it is not used as a soil-applied herbicide. Several reviews of pesticide-SOM interactions are referenced for additional reading (Walker, 1980; Weber & Miller, 1989; Weed & Weber, 1974).

A candidate pesticide is evaluated for soil bioactivity in the initial screening process. Dosage-response curves are defined as the pesticide is developed. Pesticides with soil activity are evaluated across a range of soils, crops, and pests before registration is received. Usually the pesticide label adjusts the application rate for differences in soil texture, SOM, or CEC. A three tier adjustment is common with soils listed as course, medium, or fine textured. An additional adjustment may divide soils with <30 g kg^{-1} (3%) SOM from those >30 g kg^{-1} (3%), for a six tier adjustment. Examples are metolachlor, acetochlor, and a combination of metribuzin plus chlorimurin ethyl. In a silt loam soil, the cyanazine rate increases by 1000 g kg^{-1} (100%) when SOM increases from 10 g kg^{-1} (1%) to 50 g kg^{-1} (5%). Extremes in rate adjustment variations are imazethapyr with only one rate for all soils and cyanazine with five soil texture and six SOM levels for 30 rate categories ranging from 1.3 to 5.3 kg a.i. ha^{-1}. Rate adjustments are often made for crop safety and efficacy considerations, but carryover potential, environmental concerns, and climatic conditions are also factors. Some labels adjust rates for regional differences and some have pH restrictions.

Because of the extensive work that goes into labeling a pesticide and the legal implications of not using a pesticide according to its label, seldom do soil testing laboratories or consultants suggest application rates that are not consistent with the label. All surveyed soil testing laboratories indicated pesticide suggestions were made according to labeled instructions. Accordingly, the role of a soil testing laboratory is to determine the soil texture, SOM, CEC, and pH accurately so that correct pesticide rates can be selected according to the label.

SUMMARY

In summary, 35 commercial and university laboratories were surveyed. Of this 35, one laboratory did not run SOM, one determined humic matter, 15 analyzed for SOM routinely, and 19 laboratories analyzed for SOM as an option. Of the laboratories analyzing for SOM routinely, five university and four commercial laboratories use SOM as a factor in making fertilizer N recommendations for at least corn. Wisconsin and Missouri use SOM to make fertilizer N recommendations for a wide range of grain, vegetable, and oil crops. Soil organic matter content is used to separate soils into categories for purposes of lime recommendations. All laboratories reported that their herbicide recommendations (if made) follow label suggestions.

REFERENCES

Buchholz, D.D. 1983. Soil test interpretations and recommendations handbook. Mimeo. Dep. of Agronomy, Univ. of Missouri, Columbia.

Frank, K., and D. Knudsen. 1989. Understand your soil test: Sulfur. NebGuide G89-901. Coop. Ext. Serv., Inst. of Agric. and Nat. Resour., Univ. of Nebraska, Lincoln.

Gregorich, E.G., C.M. Moreal, B.H. Ellert, D.A. Angers, and M.R. Carter. 1993. Evaluating changes in soil organic matter. p. 10-1–10-17. In D.F. Acton (ed.) A program to assess and monitor soil quality in Canada. Soil Quality evaluation program summary (interim). Land and Biol. Resour. Res. Centre. no. 93-49. Res. Branch Agric. Canada, Ottawa.

Kelling, K., E. Schulte, L. Bundy, S. Combs, and J. Peters. 1991. Soil test recommendations for field, vegetable and fruit crops. Univ. of Wisconsin Coop. Ext., Madison.

Mehring, A.L., and G.E. Bennett. 1950. Sulfur in fertilizers, manures, and soil amendments. Soil. Sci. 70:73–82.

National Academy of Science 1987. Nitrates: An environmental assessment. p. 214. Natl. Res. Coun., Washington, DC.

Penas, E., G. Hergert, and R. Ferguson. 1994. Fertilizer suggestions for corn. NebGuide G74-174-A. Coop. Ext., Inst. of Agric. and Nat. Resour., Univ. of Nebraska, Lincoln.

Rehm, G., M. Schmitt, and R. Munter. 1993. Fertilizer recommendations for agronomic crops in Minnesota. Bu-6240E. Minnesota Ext. Serv., Univ. of Minnesota, St. Paul.

Sander, D., and K. Frank. 1980. Fertilizing grain sorghum. NebGuide G74-112. Coop. Ext. Serv., Inst. of Agric. and Nat. Resour., Univ. of Nebraska, Lincoln.

Thompson, L.M., C.A. Black, and J.A. Zoellner. 1954. Occurrence and mineralization of organic phosphorus in soils, with particular reference to associations with nitrogen, carbon, and pH. Soil Sci. 77:185–196.

Tucker, R., and R. Messick. 1995. Crop fertilization based on N. C. soil tests. Circ. no. 1 (Revised 1995). North Carolina Dep. of Agric. Agronomic Div., Raleigh.

Walker, A. 1980. Activity and selectivity in the field. p. 203–222. In R.J. Hance (ed.) Interactions between herbicides and the soil. Academic Press, New York.

Weber, J.B., and C.T. Miller. 1989. Organic chemical movement over and through soil. p. 305–334. In B.L. Sawhney and K. Brown (ed.) Reactions and movement of organic chemicals in soils. SSSA Spec. Publ. 22. SSSA and ASA, Madison, WI.

Weed, S.B., and J.B. Weber. 1974. Pesticide-organic matter interactions. p. 39–66. In W.D. Guenzi (ed.) Pesticides in soil and water. SSSA, Madison, WI.

5 Assessing Soil Quality by Testing Organic Matter

Lawremce J. Sikora and Vladimir Yakovchenko

USDA-ARS Beltsville Agricultural Research Center
Beltsville, Maryland

Cynthia A. Cambardella

National Soil Tilth Laboratory
Ames, Iowa

John W. Doran

Soil and Water Conservation Research Unit
Lincoln, Nebraska

Soil is a finite resource and interest in measuring the degradation, improvement, or preservation of soil has led to efforts in developing tests to evaluate the quality of soil. Larson and Pierce (1991) defined soil quality as its physical, biological and chemical properties that (i) provide a medium for plant growth, (ii) regulate and partition water flow in the environment, and (iii) serve as an environmental buffer in the formation, attenuation, and degradation of environmentally hazardous compounds. Doran and Parkin (1994) defined soil quality as the capacity of a soil to function within ecosystem boundaries, to sustain biological productivity, maintain environmental quality and promote plant and animal health. Parr et al. (1992) stated that soil quality should serve as an indicator of change in both the soil's ability to produce optimum levels of safe and nutritious food, and its structural and biological integrity, which in turn is related to the status of certain degradative processes and to environmental and biological plant stress.

Soil quality traditionally has focused on, and has been equated with, agricultural system productivity or, more simply, system productivity. Crop yield is an important indicator of system productivity that is, in part, dependent upon soil quality. Crop yield can serve as a bioassay for several interacting factors such as soil, water, air, disease, germplasm, and management. Crop yield alone, however, is an incomplete measure of system productivity. Soil quality may better represents system productivity and function.

SOIL ORGANIC MATTER AND SOIL QUALITY

Soil organic matter (SOM) is considered the single most important indicator of soil quality (Larson & Pierce, 1991). Doran and Parkin (1994) have includ-

---

Copyright © 1996. Soil Science Society of America, 677 S. Segoe Rd., Madison, WI 53711, USA.
Soil Organic Matter: Analysis and Interpretation. SSSA Special Publication no. 46.

Table 5–1. Basic soil physical, chemical, and biological indicators for screening soil quality and their relationship to soil organic matter (SOM; after Doran & Parkin, 1994).

Soil characteristic	Relationship to soil condition–function	Relation to soil organic matter
Physical indicators		
Texture	Retention and transport of water and chemicals	Determines degree of SOM protection and equilibrium level
Topsoil and rooting depth	Estimate of productivity potential and erosion	Correlated with SOM
Soil bulk density and infiltration	Indicators of compaction and potential for leaching productivity and erosivity	Correlated with SOM
Water holding capacity	Related to water retention, transport, and erosivity	Correlated with SOM
Temperature	Determines plant productivity, microbial activity, and SOM level	Related to soil color and SOM
Chemical indicators		
Total organic C and N (SOM)	Defines soil fertility, stability, and erosion extent	Stability of SOM related to C/N ratio
Electrical conductivity	Defines plant and microbial activity thresholds	Effect varies with SOM content
pH	Defines biological and chemical activity thresholds	Stability and activity of SOM fractions
Cation-exchange capacity	Defines equilibrium levels of cation nutrients and H^+	Correlated with SOM and clay content
Extractable N, P,	Productivity and potential N loss indicators	Influenced by SOM transformations
Biological indicators		
Microbial biomass C and N	Flux of nutrients and pool of active N and C	Early warning indicator of SOM change
Potentially mineralizable N	Soil productivity and N supplying potential	Active SOM pool
Soil respiration	Indicator of biomass activity	Indicator of SOM turnover, early warning indicator of SOM change

ed SOM as a major component of a required minimum data set for screening soil quality (Table 5–1). The effects of management practice on SOM are critical to evaluating the sustainability of cropping and tillage systems, and concomitantly, their effects on the environment.

The Soil Science Society of America defines SOM as the organic fraction of soil exclusive of undecayed plant and animal residue (SSSA, 1987). The term *humus* is often used synonymously with SOM. This definition, however, seems too restrictive given the fact that most analytical procedures for SOM do not distinguish between decomposed and undecomposed plant and animal residues and living organisms in soils sieved to <2 mm before analysis. In the broadest sense, SOM consists of diverse components such as living organisms, slightly altered plant and animal organic residues, and well-decomposed organic residues that vary considerably in their stability and susceptibility to further decomposition (Magdoff, 1992). These SOM components contain up to 99% of the total N in soil.

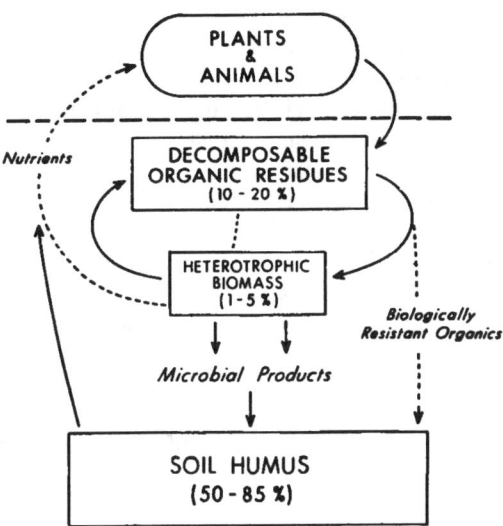

Fig. 5–1. Relationships between living and nonliving soil organic matter components, their role in nutrient cycling, and their relative proportion of total soil organic matter (after Doran & Smith, 1987).

Interpretation and prediction of the effects of soil management on N availability to plants, or loss to the environment, depend on understanding the unique roles played by living and nonliving components of SOM. As illustrated in Fig. 5–1, the majority of SOM is contained in plant and animal debris and soil humus. These nonliving components of SOM play an important role in determining the soil physical–chemical environment within which living organisms function. Heterotrophic soil microorganisms and fauna, a relatively small proportion of total OM (1–5%), function as important catalysts for transformation and cycling of N and other nutrients. The importance of soil microbial biomass as a significant sink–source for C and N is discussed by Jenkinson (1988).

Soil organic matter is associated with different size soil particles. Dalal and Mayer (1986) found that after cultivation organic C associated with sand-sized particles declined rapidly; that associated with silt-sized particles declined the least; and that associated with clay-sized particles increased from 48 to 61%. These data suggest that clay particles protect SOM. Bell (1993) referred to another division of SOM, namely soils having *active* and *passive* OM. Similarly, Sikora and McCoy (1990) described efforts to measure C in soils that is similar to the active fraction of SOM.

Grassland soils traditionally lose SOM rapidly after they are first cultivated (Allison, 1973). The loss of SOM is usually exponential, declining rapidly during the first 10 to 20 yr, then continuing more slowly until a new equilibrium is reached after 50 to 60 yr (Jenny, 1941). The SOM level declines to a new steady-state level regulated by newly established abiotic and biotic parameters of the cropped ecosystem (Tate, 1987). The reasons for SOM decline are stimulation of decomposition after plowing or cultivation and a decrease in plant residue additions. Therefore, any factor that increases the plant biomass or slows the decomposition of SOM will tend to increase the equilibrium level of SOM.

Soil texture and soil environmental conditions affect SOM decomposition rates. Spain (1990) and Feller et al. (1991) found significant positive correlations between soil texture and soil organic C content. The protection of SOM from decomposition is attributed to adsorption of SOM onto clay surfaces (Oades, 1989), encapsulation by clay particles (Tisdall & Oades, 1982) or entrapment of SOM in small pores of aggregates making SOM inaccessible to microorganisms (Elliot & Coleman, 1988). Both temperature and moisture regimes affect the equilibrium SOM content of soils (Buckman & Brady, 1969). Increased temperature decreases SOM content, while increased moisture increases SOM. Within belts of uniform moisture conditions and comparable vegetation, average total SOM and N contents increase two to three times for each 10°C decline in mean annual temperature. With all these factors taken into consideration, it is understandable that no SOM standard exists that would designate a level of soil quality.

SOIL ORGANIC MATTER LEVELS

The determination of soil quality by measuring SOM requires that the quantity of SOM be compared with a standard. The question is what should the standard quantity be for soils considered to have good quality? Can quantity in fact be equated to quality? When measuring SOM, is quantity, content, or change in SOM quantity or content important to consider in determining soil quality? Larson and Pierce (1994) suggest monitoring the dynamics of SOM change. Sustainable agriculture principles suggest that change be minimized as long as the management practice is productive and there is a balance of inputs and outputs in the system. If SOM changes little with time, and the system is productive and resilient, the SOM content may be considered optimum. Buckman and Brady (1969) suggest that SOM should be maintained at an economic maximum consistent with a suitable physical condition of the soil, satisfactory biochemical activity, adequate availability of nutrients, and profitable crop yields. Essentially, there is not an established quantity of SOM that can be defined as *good* or *less than optimum* quality because SOM evaluation must include farming practice and climate as covariables within soil texture and drainage classes.

Farming practices can significantly influence SOM content but the changes are slow in temperate climates (Johnston, 1993). In Missouri, 50 yr of cropping to continuous corn (*Zea mays* L.) caused a 56% decline in SOM as compared with SOM levels of virgin timber or mixed grasses (Table 5–2). Continuous cropping to timothy (*Phleum pratense* L.) caused the least decline in SOM (19%) and 3- to 6-yr crop rotations were intermediate in SOM loss. The major influence of crop management practice on SOM levels was likely related to differences in the amount of crop biomass produced that was partitioned above and belowground as reflected by abundance and nature of crop rooting characteristics. Even with continuous corn, however, where SOM losses were greatest, it would take 3 to 5 yr before the decline in total SOM levels was detectable by conventional analytical procedures. This study illustrates the importance of understanding the potential

Table 5-2. Soil organic matter (SOM) content recorded after 50 yr of crop rotations at the Sanborn field (from USDA, 1980). Data are from single plots established in 1888.

Crop rotations	Soil organic matter content
Continuous corn	1.45†
Continuous wheat	3.4
Continuous oats	4.08
Continuous timothy	4.68
Corn, wheat, red clover rotation	3.31
Corn, oats, wheat, red clover rotation	3.74
Corn, oats, wheat, red clover, timothy, timothy rotation	3.83
Virgin grass and timber	5.78

† Converted from N to SOM by multiplying by 17.

equilibrium levels of SOM that are attainable using different cropping systems and the need for more sensitive short-term or early warning indicators of long-term trends in SOM change.

Recent concern for increasing atmospheric CO_2 levels and global warming has resulted in increased interest in the sequestration or tie-up of plant C by SOM. Long-term crop rotations and N fertilizer practices can result in higher equilibrium SOM contents due to greater residue additions and/or lower decomposition rates. Results of an 8-yr study in the Western corn belt demonstrate that from 100 to 200 kg of C ha^{-1} yr^{-1} can be sequestered by SOM, even in some continuous cropping systems that receive sufficient levels of N fertilizer (Table 5-3). Although statistically significant, these increases in SOM represent <3% of the total C present in soil. This research illustrates that increasing the SOM content of temperate area soils is a slow process. Practical evaluation of SOM change as an indicator of soil quality will require specific analysis of the dynamic SOM fractions that define SOM quality and serve as indicators of potential long-term change in the quantity of SOM.

Table 5-3. Crop rotation and fertilization effects on C sequestration in tilled soil layer (0–15 cm) in an 8-yr study conducted in Mead, NE (Varvel, 1994). Orthogonal comparisons indicated significant increases in organic C in the fertilized vs. unfertilized plots and 4-yr vs. 2-yr rotations in the 0- to 7.5-cm depth, but no significant difference in the 7.5- to 15-cm depth.

Cropping practice	Relative N fertilizer rate		
	None	Low	High
	kg C ha-1 8 yr^{-1}		
	Continuous		
Corn (c)	−534	186	1358
Soybean (sb)	−855	−203	−363
Sorghum (sg)	259	1316	1538
	Rotations		
c–sb	−283	186	−69
sg–sb	333	228	892
c–ocl–sb–sb†	870	644	1420
c–sb–sg–ocl	454	1012	1261

† ocl, oat and clover.

BIOLOGICAL INDICATORS

Efforts to improve our ability to predict and manage soil quality depend to a large extent on increasing our understanding of the processes that control SOM turnover. The biological factors associated with organic C are microbial biomass, respiration, enzymes and active C fractions such as carbohydrates and light fraction (Gregorich & Ellert, 1993). These factors tend to change more dramatically than total SOM and could be valuable indicators of soil quality or changes in quality. Yakovchenko et al. (1996) evaluated several biological factors in soils from the Rodale Institute Research Center Farming Systems Trial. The trial compared conventional farming to low-input cash grain or low-input legume practices. Low-input cash grain obtained N from manures or legumes and low-input legume obtained N solely from legumes. Although the change of total organic C is not different from the conventional systems, the biological or labile fractions in the low-input systems are significantly different (Table 5–4). These SOM components or fractions may serve as early warning indicators of quality; the interpretation of these data, however, are difficult because yields or crop productivity have not been different (Yakovchenko et al., 1996).

Opinions differ as to whether microbial biomass can act as an indicator of soil quality. Campbell et al. (1991) suggest that an increase in biomass N does not necessarily denote an improvement in soil quality. McGill et al. (1986) reported that there was a highly significant correlation between average biomass C and total crop yields during 5 yr. ($R = 0.80$, $P < .001$). Biederbeck et al. (1984) found more biomass C and N in continuous wheat (*Triticum aestivum* L.) receiving no fertilizer as compared with continuous wheat receiving fertilizer N. They suggested that there was a larger but less active microbial population present in the poorly fertilized system. Carbon dioxide evolution and microbial respiration/biomass C ratio as well do not necessarily correlate to soil productivity. Dinwoodie and Juma (1988) reported that more C was lost (during 10-d incubation) by

Table 5–4. Mean and standard error of total organic C, particulate organic matter, biomass C, and specific respiration (CO_2–C Biomass C^{-1}) of 0- to 20-cm surface soil from the Farming System Trial, 1992 (Yakovchenko et al., 1996).

Treatment	Total organic C	POM†	Biomass C	Biomass C total organic C^{-1}	CO_2–C‡	CO_2–C biomass C^{-1}
	Mg ha^{-1}	% of total C	kg ha^{-1}	%	kg ha^{-1} d^{-1}	d^{-1}
Low input animal	46 ± 2	20.2 ± 2.1	410 ± 38	0.89 ± 0.09	19.8 ± 2.2	0.048 ± 0.007
Low input legume	49 ± 2	20.1 ± 2.3	298 ± 24	0.61 ± 0.05	15.7 ± 1.1	0.052 ± 0.008
Convent cash grain	46 ± 1	15.7 ± 1.4	250 ± 20	0.54 ± 0.04	9.7 ± 0.6	0.039 ± 0.004
Grass border	59 ± 3	ND§	698 ± 65	1.21 ± 0.12	48.0 ± 5.3	0.063 ± 0.009

† Particulate organic matter (POM) data taken from 1993 sampling and expressed as a percentage of total C in 1993 samples.
‡ Laboratory incubations at 25°C and field water content (bulk density ≈1.3).
§ Not determined.

microbial respiration from less productive than more productive soils. Carbon dioxide–C evolution was up to two-fold greater when expressed on an area basis (mg m^{-2}), and 2- to 5-fold greater when expressed as milligram per gram of microbial C. Chien et al. (1964), on the contrary, reported that respiratory capacity increased with improvement in soil fertility. These data suggest that biomass, CO_2 evolution, and specific respiration changes are much larger than SOM changes and that these fractions may allow more rapid evaluation of treatment effects.

ORGANIC MATTER FRACTIONS AS INDICATORS

The use of SOM as an indicator of quality must focus on the discrete pools even though the ultimate concern is with the loss of total SOM. Only the most labile forms of SOM are thought to be involved in supplying nutrients for plant growth; however, the most labile fractions are also the first be depleted as a result of cultivation or other system perturbations (Doran & Smith, 1987). The dynamics of intermediately labile SOM, a pool that turns over every 15 to 20 yr, are most severely affected by these disturbances. Therefore, increasing SOM storage through inputs to intermediately-labile SOM pools may be a requirement for system long-term sustainability.

Current research suggests that characteristics of particulate organic matter (POM), an analytical fraction obtained by physical fractionation of soil, are consistent with theoretical characteristics of intermediately labile SOM pools. In native grassland soils, POM C can account for up to 48% of the total soil C and 32% of the total soil N (Greenland & Ford, 1964). Thirty-nine percent of the total soil organic C from the top 20 cm of a native grassland soil in Western Nebraska is associated with the POM fraction (Cambardella & Elliott, 1992). Similar results were reported for Canadian native prairie soils by Tiessen and Stewart (1983). The POM fraction consists of partially decomposed pieces of plant residue and has a density of <1.85 g cm^{-3} and a C/N ratio of 20. Light fraction (LF) SOM, as defined originally by Greenland and Ford (1964), has a similar density (<2.0 g cm^{-3}) and C/N ratio (<25). For the purposes of this discussion, we consider LF SOM to be analogous to POM.

The POM shows promise as a short-term or early warning indicator of long-term changes in soil quality. The POM has been shown to be much more responsive than total SOM to changes in agricultural management (Paustian et al., 1993) and, as such, has been suggested an indicator of soil quality (Gregorich & Ellert, 1993). Greenland and Ford (1964) reported that grassland POM C concentrations were reduced by a factor of four after cultivation. For a long-term tillage experiment located in the vicinity of Saskatchewan, Canada, the percentage of total soil C associated with POM was reduced by 65% as a result of 60 yr of wheat–fallow cultivation (Tiessen & Stewart, 1983). Cambardella and Elliott (1992) reported that >50% of the organic C in the POM fraction is lost as a result of 20 yr of plowing and bare fallow management. Continuous cropping to cereal grains for 70 yr resulted in an average loss of >50% of the organic C from the POM fraction for the six major soil series in Southern Queensland (Dalal & Mayer, 1986). These studies clearly demonstrate that cultivation results in the selective loss of SOM from the POM fraction, and that much of this loss occurs relatively quickly (with 4–10 yr) after the initiation of cultivation.

Certain agricultural management practices, such as reduced fallow frequency, reduced tillage, seeding to grasses or legumes, and prudent manure and fertilizer application can ameliorate the detrimental effects of cultivation and encourage POM C and N to accumulate. Greenland and Ford (1964) observed lower amounts of POM C and N with wheat–fallow management compared with continuous wheat. The POM C was nearly 50% higher for soils under no-till compared with plowing for a wheat–fallow system in western Nebraska (Cambardella & Elliott, 1992). No-tillage increased POM C by 20% compared with conventional tillage for a grain sorghum–winter rye cropping system near Athens, GA (Beare et al.,1994). When aboveground crop residue is removed during silage corn production, reduced tillage still increased POM C and N compared with plowing, suggesting the importance of root inputs into this fraction (Angers et al., 1993). Increases in POM C and N associated with no-tillage have been correlated with increases in macroaggregation and with increased amounts of C and N in macroaggregates (Cambardella & Elliott, 1993; Beare et al., 1994). Data collected from Rodale's long-term Farming Systems Trial experiment demonstrated that for organic management systems, a cover-crop (mixed legume–small grain) based system accumulated the highest amount of C and N in the POM fraction compared with a manure-based system and a conventional cash–grain system. The manure based system had higher POM C and N than the conventional system, but the trend was not statistically significant (Wander et al., 1994). For three long-term study sites in western Canada situated along an evapotranspiration gradient, POM C and N were greatest for soils that were continuously cropped to spring wheat and for cropping systems with reduced fallow frequency that included alfalfa hay in the rotation (Janzen et al., 1992). Nutrient amendments (Biederbeck et al., 1994) and manure application (Bremer et al., 1994) enhanced the beneficial effect of reduced fallow on POM C and N accumulation for wheat-based cropping systems.

These studies confirm the usefulness of the POM fraction as a sensitive early indicator of long-term management induced changes in SOM. Recent research has found that >50% of the N contained in the POM fraction was lost between April and the middle of August under a no-till corn–soybean cropping system, establishing a possible link between N losses from the POM fraction and nutrient supply over the growing season (Cambardella, 1994). Yakovchenko and others (1996, unpublished data), showed that while more C was evolved as CO_2 in a soil amended with additional POM, N mineralization was not increased. These data suggest that POM increase may not directly affect nutrient availability; however, the changes in POM that often occur prior to SOM changes suggest that the intermediate C pool which is represented by POM may be a soil quality indicator that reflects the potential changes in SOM.

CONCLUSIONS

Soil organic matter is a major component in the evaluation of soil quality. Determination of total SOM is informative when farming practice and climate are considered as covariables. Because SOM content changes slowly in temperate climates, determination of a labile portion of SOM may be more indicative of treatment effects. Suggestions of SOM components that may be useful in soil quality evaluation and serve as early warning indicators are particulate organic matter or POM, microbial biomass and specific respiration.

REFERENCES

Allison, F.E. 1973. Soil organic matter and its role in crop production. Devel. in Soil Sci. 3. Elsevier, Amsterdam.
Angers, D.A., A. N'dayegamiye, and D. Cote. 1993. Tillage-induced differences in organic matter of particle-size fractions and microbial biomass. Soil Sci. Soc. Am. J. 57:512–516.
Beare, M.H., P.F. Hendrix, and D.C. Coleman. 1994. Water-stable aggregates and organic matter fractions in conventional and no-tillage soils. Soil Sci. Soc. Am. J. 58:777–786.
Bell, M.A. 1993. Organic matter, soil properties and wheat production in the high valley of Mexico. Soil Sci. 156:86–93.
Biederbeck V.O., C.A. Campbell, and R.P. Zentner. 1984. Effect of crop rotation and fertilization on some biological properties of a loam in South Western Saskatchewan. Can. J. Soil Sci. 64:355–367.
Biederbeck, V.O., H.H. Janzen, C.A. Campbell, and R.P. Zentner. 1994. Labile soil organic matter as influenced by cropping practices in an arid environment. Soil Biol. Biochem. 26:1647–1656.
Bremer, E., H.H. Janzen, and A.M. Johnston. 1994. Sensitivity of total, light fraction and mineralizable organic matter to management practices in a Lethbridge soil. Can. J. Soil Sci.74:131–138.
Buckman, H.O., and N.C. Brady. 1969. The nature and properties of soils. Macmillian Co., London.
Cambardella, C.A. 1994. Temporal dynamics of particulate organic matter N and soil nitrate N with and without an oat cover crop. Bull. Ecol. Soc. Am.. 75:30.
Cambardella, C.A., and E.T. Elliott. 1992. Particulate soil organic matter changes across a grassland cultivation sequence. Soil Sci. Soc. Am. J. 56:777–783.
Cambardella, C.A., and E.T. Elliott. 1993. Carbon and nitrogen distribution in aggregates from cultivated and native grassland soils. Soil Sci. Soc. Am. J. 57:1071–1076.
Campbell C.A., V.O. Biederbeck, R.P. Zentner, and G.P. Lafond 1991. Effect of crop rotations and cultural practices on soil organic matter, microbial biomass and respiration in thin black chernozem. Can. J. Soil Sci. 71:363–376.
Chien T., F. Ho, H. Feng, S. Liu, and P. Chen. 1964. The microbial characteristics of red soils. Acta Pedol. Sin. 12:399–400.
Dalal, R.C., and R.J. Mayer. 1986. Long-term trends in fertility of soils under continuous cultivation and cereal cropping in southern Queensland: III. Distribution and kinetics of soil organic carbon in particle-size fractions. Aust. J. Soil Res. 24:293–300.
Dinwoodie, G.D., and N.G. Juma. 1988. Allocation and microbial utilization of C in two soils cropped to barley. Can. J. Soil Sci. 68:495–505.
Doran, J.W., and T.B. Parkin. 1994. Defining and assessing soil quality. p. 3–21. *In* J.W. Doran et al. (ed.) Defining soil quality for a sustainable environment. SSSA Spec. Publ. 35. SSSA and ASA, Madison, WI.
Doran, J.W., and M.S. Smith. 1987. Organic matter management and utilization of soil and fertilizer nutrients. p. 53–72. *In* R.F. Folett et al. (ed.) Soil fertility and organic matter as critical components of production systems. ASA Spec. Publ. 19. ASA and SSSA, Madison, WI.
Elliot, E.T., and D.C. Coleman. 1988. Let the soil work for us. Ecol. Bull. 39:23–32.
Feller, C., E. Fritsch, R. Ross, and C. Valentin. 1991. Effects of the texture on the storage and dynamics of organic matter in some low activity clay soils (West Africa, particularly). Cah. ORSTOM, Ser. Pedol. 26:25–36.
Greenland, D.J., and G.W. Ford. 1964. Separation of partially humified organic materials from soils by ultrasonicdispersion. p. 137–148. *In* Trans. Int. Congr. Soil Sci. 8th, Bucharest. 31 Aug.–9 Sept. 1964. Rompresfilatelia, Bucharest.
Gregorich, E.G., and B.H. Ellert. 1993. Light fraction and macroorganic matter in mineral soils. p. 397–407. *In* M.R. Carter (ed.) Soil sampling methods of analysis. Lewis Publ., Boca Raton, FL.

Janzen, H.H., C.A. Campbell, S.A. Brandt, G.P. Lafond, and L. Townley-Smith. 1992. Light-fraction organic matter in soils from long-term crop rotations. Soil Sci. Soc. Am. J. 56:1799–1806.

Jenkinson, D.S. 1988. Determination of microbial biomass carbon and nitrogen in soil. p. 368–386. *In* J.R. Wilson (ed) Advances in nitrogen cycling in agricultural ecosystems. CAB Int., Oxon, England.

Jenny, H. 1941. Factors of soil formation. McGraw-Hill, New York.

Johnston, A.E. 1993. Significance of organic matter in agricultural soils. p. 1–18. *In* Organic substances in soil and water: Natural constituents and their influences on contaminant behaviour. Spec. Publ. 135. Royal Soc. of Chem., Cambridge, England.

Larson, W.E., and F.J. Pierce. 1991. Conservation and enhancement of soil quality. p. 175–203. In Evaluation of sustainable management in the developing world. Vol 2. IBSRAM Proc. 121(2). Thailand Int. Board for Soil Res. and Management, Bangkok.

Larson, W.E., and F.J. Pierce. 1994. The dynamics of soil quality as a measure of sustainable management. p. 37–51. *In* J.W. Doran et al. (ed.) Defining soil quality for a sustainable environment. SSSA Spec. Publ. 35. SSSA and ASA, Madison, WI.

Magdoff, F. 1992. Building soils for better crops: Organic matter management. Univ. of Nebraska Press, Lincoln.

McGill, W.B., K.R. Cannon, J.A. Robertson, and F.D. Cook. 1986. Dynamics of soil microbial biomass and water-soluble organic C in Breton L after 50 years of cropping to two rotations. Can. J. Soil Sci. 66:1–19.

Oades, J.M. 1989. An introduction to organic matter in mineral soils. p. 89–159. *In* J.B. Dixon and S.B. Weed (ed.) Minerals in soil environments. Soil Sci. Book Ser. 1. SSSA, Madison, WI.

Parr, J.F., R.I. Papendick, S.B. Hornick, and M.E. Meyer. 1992. Soil quality: Attributes and relationship to alternative and sustainable agriculture. Am. J. Altern. Agric. 7:5–11.

Paustian, K.E., T. Elliott, H.P. Collins, C.V. Cole, and E.A. Paul. 1993. Changes in active C fractions as a function of land management practices. p. 257. *In* Agronomy abstracts. ASA, Madison, WI.

Sikora, L.J., and J.L. McCoy. 1990. Attempts to determine available carbon in soil. Biol. Fert. Soils. 9:19–24.

Soil Science Society of America. 1987. Glossary of soil science terms. SSSA, Madison, WI.

Spain, A. 1990. Influence of environmental conditions and some soil chemical properties on the carbon and nitrogen contents of some tropical Australian rain-forest soils. Aust. J. Soil Sci. Res. 28:825–839.

Tate, R.L., III. 1987. Soil organic matter: Biological and ecological effects. John Wiley & Sons, New York.

Tiessen, H., and J.W.B. Stewart. 1983. Particle-size fractions and their use in studies of soil organic matter: II. Cultivation effects on organic matter composition in size fractions. Soil Sci. Soc. Am. J. 47:509–514.

Tisdall, J.M., and J.M. Oades. 1982. Organic matter and water stable aggregates in soil. J. Soil Sci. 33:141–163.

U.S. Department of Agriculture. 1980. Report and recommendations on organic farming. U.S. Gov. Print. Office, Washington, DC.

Varvel, G.E. 1994. Rotation and N fertilizer effects on changes in soil carbon and nitrogen. Agron. J. 86:319–325.

Wander, M.M., S.J. Traina, B.R. Stinner, and S.E. Peters. 1994. Organic and conventional management effects on biologically active soil organic matter pools. Soil Sci. Soc. Am. J. 58:1130–1139.

Yakovchenko, V., L.J. Sikora, and D.D. Kaufman. 1996. A biologically-based indicator of soil quality. Biol. Fert. Soils 21:245–251.

6 Carbon Fractions in Compost and Compost Maturity Tests

Charles L. Henry and Robert B. Harrison
University of Washington
Seattle, Washington

Composting is generally considered as the controlled biological degradation of organic materials. It is undoubtedly both an aerobic and an anaerobic process. Although probably aerobic by design, in many cases, composting operations are run in anaerobic conditions either on purpose or inadvertently, and there also exist many anaerobic microsites within the compost pile. During this decomposition process, organic compounds decompose at different rates; loosely defining stages of composting as shown in Fig. 6–1. The readily decomposed organics (i.e., sugars, starches, fats, and proteins) breakdown during the rapid composting period in the first few weeks; the organics slower to degrade (i.e., hemicellulose and cellulose), do so during the curing or maturing stage of composting; while lignin and lignocellulose are usually not decomposed significantly during composting. Maturity may be defined as the point at which the compost will not act detrimentally when used as a soil amendment. In other words, the rate of decomposition of the remaining organic components does not cause anaerobic conditions and subsequent odors, and produces acceptably small amounts of organic acids (potentially toxic to plants).

Much, or in some cases, all of the feedstock used in composting is organic matter derived from the cellular material of plants and animals. This organic matter can roughly be fractionated similar to the categories shown in Fig. 6–1 by a sequential reflux–ashing method modified from Van Soest (1963). We used this analytical technique to give us some preliminary data on measures of: (i) the non-cell wall mass, (ii) cellulose and hemicellulose, (iii) lignin, and (iv) ash. Figure 6–2 shows data on variety of compost feedstock and biosolids according to these categories.

CHANGES IN CARBON FRACTIONS DURING COMPOSTING

The dominant element in the decomposition process is C. Figure 6–3 describes the potential paths of the C cycle in a compost pile. Complete oxidation of organic matter will yield CO_2, H_2O, and minerals; however, many intermediaries will be formed—some unstable, some stable—which either continue to decompose at a slower rate, or contribute to what may be considered a final compost product. Composting is a relatively short process compared with the time it

---

Copyright © 1996. Soil Science Society of America, 677 S. Segoe Rd., Madison, WI 53711, USA.
Soil Organic Matter: Analysis and Interpretation. SSSA Special Publication no. 46.

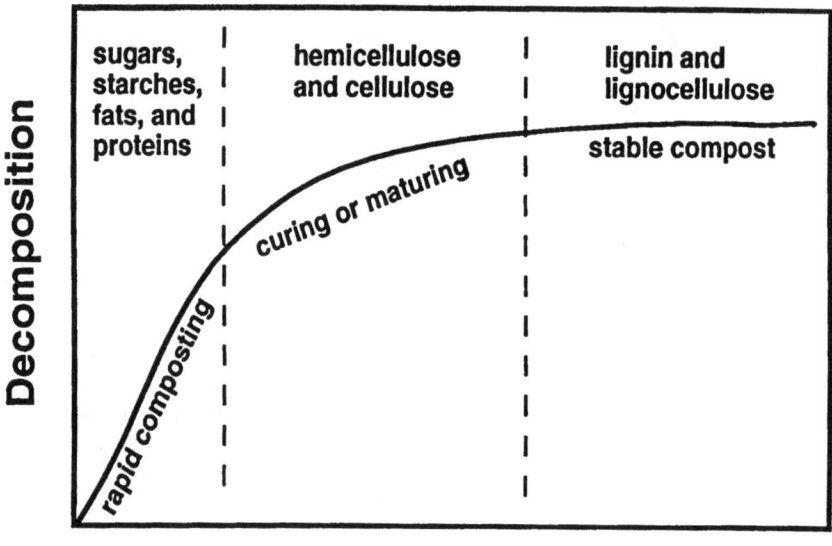

Fig. 6–1. Conceptualized rate of decomposition of organic compounds, loosely defining stages of composting.

takes for soil humus to form, and thus, in addition to significant amount of humic (HA) and fulvic acids (FA) in lignin-humus complexes along with ash, there also will remain more recalcitrant carbohydrates, and some amount of microbial biomass.

As shown in Fig. 6–1, composting proceeds in somewhat defined stages. The first stage is the rapid decomposition of high energy organics like simple carbohydrates and proteins primarily by bacteria. Due to this energy released in the initial stage, the pile heats up and correspondingly increases microbial activity, until and unless temperatures rise to the point of inactivating microbes. This stage is relatively short, primarily due to the rate of decomposition, and depletion of this easily decomposed substrate.

During the second stage of composting (after the first 2–6 wk), the microbial population shifts to those that decompose cellulose and hemicellulose; primarily the actinomicetes and fungi. These organic compounds are harder to decompose and yield less energy, thus the rate of decomposition is reduced and less heat is given off. This stage is usually considered the curing or maturing stage, and can take from 2 to 12 mo. Beyond this stage, little decomposition occurs during what we consider composting.

Curing is the stage that leads to maturity. As organics convert more and more towards humus, they correspondingly become more stable. Because we have defined maturity as the point at which the compost will not act detrimentally when used as a soil amendment, end uses require different curing times. For instance, a compost used out in a field where it will sit for a period of months without planting, and where odors are not a problem may need little if any curing

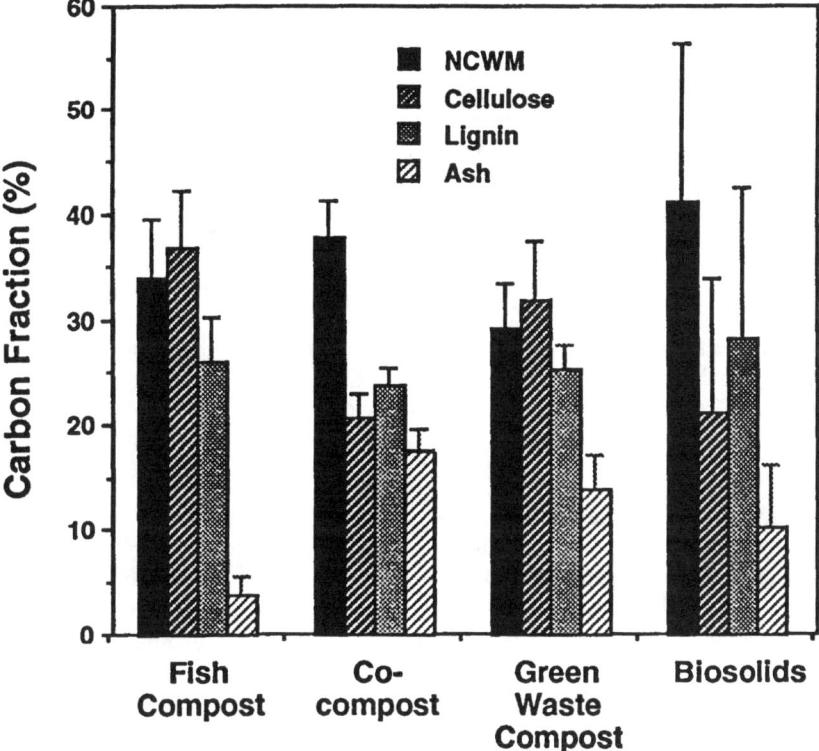

Fig. 6–2. Fractions of C compounds in a number of composts analyzed by the University of Washington, including the noncell wall mass, cellulose, lignin, and ash.

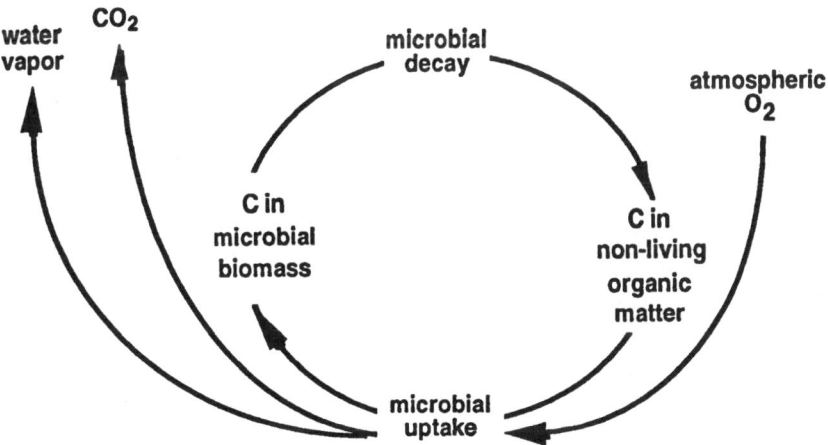

Fig. 6–3. Simplified C cycle as it relates to aerobic composting.

time to become mature for our use. Conversely, a bagged compost will require a stable organic mass so that odors or mycelium growth are not present, discouraging the consumer.

MEASUREMENTS OF MATURITY

There is no test for compost stability that proves consistent for all composts, without regard to parent material. Many tests are relative tests, i.e., the feedstock or the intermediate composted material is compared with a final product. These tests include: (i) biological, (ii) chemical, (iii) microbial, and (iv) physical indicators. Much of the following discussion of these tests was extracted from a literature review performed for the Washington State Department of Ecology and updated for the Solid Waste Compost Council authored by M. Knoop, K. Talbott, H. Wescott, and C. Henry (Henry, 1991).

Biological Indicators

Plant bioassays are perhaps one of the most appropriate tests for maturity because they indicate when a compost may be used without inhibitory effects. They are slow, taking from 2 to 3 wk for results. Yield reductions of komatsuna (*Brassica rapa* L.H. Bailey var. *pervidis*) up to 50% compared with controls were found by Chanyasak et al. (1983a,b) from a 10 dry t ha^{-1} loading of immature municipal solid waste (MSW) composts, especially in early growth stages. Growth in mature compost was similar or slightly higher than controls. High loading of both mature and immature composts (20 dry t ha^{-1}) resulted in lower yields compared with control.

Phytotoxicity

A compost intended for use as a soil amendment should be mature enough to avoid phytotoxic effects while allowing for advanced stages of decomposition to provide energy within the growing medium (Inoko, 1985). Composts can be cured beyond the point of effectiveness for disease suppression (Hoitink & Fahy, 1986). Assays can determine suitability of compost by indicating cellulose content, cation-exchange capacity (CEC), C/N ratio, C_{org}/N_{org} ratio in water extracts and indicator microorganisms.

Phytotoxic response in cress (*Lepidium sativum* L.) disappeared in 47 d-old compost, while 76 d-old compost exhibited a growth stimulatory effect compared with a nursery mix (Hardy & Sivasithamparam, 1989). After an initially high CO_2 evolution, CO_2 fell steeply by Day 9, increased slowly with time and peaked at Day 243, during high temperature periods (45–55°C). Phytotoxicity was eliminated during the initial high-temperature phase. Cress experienced some growth stimulation from the compost extract after the initial thermophilic stage. The eucalyptus (*Eucalyptus calophylla* R. Br. ex Lindl. and *E. divversicolor* F.J. Muell.) bark mix required 14 mo of composting, in spite of N addition and control of moisture and temperature.

Wong (1985) found that compost toxicity decreased with age while metal concentrations remained essentially unchanged. This suggested that metals are not the main cause of toxic response in bioassays, even though certain metals at low concentrations may slow germination and hinder root growth. The decline in toxicity with age of the compost is most likely due to degradable organic compounds. Wong (1985) linked inhibition of germination and root growth to ethylene oxide and ammonia contents. Chanyasak et al. (1983a,b) attributed inhibitory effects on komatsuna growth at a low loading of MSW compost to low fatty acids, especially propionic and *n*-butyric acids. Inhibitory effects at a high loading were caused by high sodium chloride levels.

Mineral N content in an extract increased with the composting period (Juste et al., 1987). The phytotoxic effect of immature compost was related to immobility of mineral N in the material. Accordingly, N uptake was enhanced in test plants of 180- or 240-d-old composts. Nitrogen starvation also was found by Inoko (1985) to cause crop damage after addition of immature cellulose-rich compost because the carbohydrates had not been sufficiently reduced.

Germination tests for phytotoxicity, C/N, and CEC were used to verify maturity in fish waste composts (Mathur et al., 1986). Results of a germination index (phytotoxicity disappeared within 30 d) and disappearance of the molecular weight component (after 20 d) were suggested instead of other parameters of maturity because equilibrium was reached (Saviozzi et al., 1987). A germination index higher than 80 to 85% indicated the absence of phytotoxicity in 2-mo old compost (Riffaldi et al., 1986). Phytotoxic effects disappeared 30 d into the composting process. Zucconi et al. (1981) found that a highly sensitive germination index was obtained by multiplying germination and root growth. Low toxicity (levels as low as 25 ppb of compounds associated with decomposing organic matter) affected root growth; high toxicity inhibited germination. The method produced concentration-dependent responses and differentiated among stages of maturity. Stimulatory effects of mature composts were also recorded.

Seed germination of flowering Chinese cabbage (*B. parachinensis* Bailey) in MSW compost samples was lower than in the controls (Wong, 1985). High growth suppression was recorded in fresh and 2-wk old compost; 3-wk old compost produced 80% germination. Zero to 4-wk old compost produced much shorter root lengths than the control and inhibitory response was reported in fresh, 2-wk old and 3-wk old composts, becoming less in 6-wk old compost. Root lengths in compost >8-wk old were similar to controls. A curing period of at least 4 wk was necessary to reduce phytotoxicity. Root length was better than seed germination for examining toxicity effects and was suggested by Wong and Chu (1985) to indicate maturity. Tomato [*Lycopersicon lycopersicum* (L.) Karsten] was most sensitive, followed by Chinese white cabbage (*B. chinensis* L.) and carrot (*Daucus carota* Hoffm.) in seed germination and root growth tests. Root growth of flowering Chinese cabbage proved an effective and simple for determining maturity of MSW composts.

Chemical Parameters

More & Sana (1987) suggested that chemical parameters are useful in determining maturity in composts but dependence on any one will provide only partial information. Various trends of chemical parameters during composting have been reported but no single test is consistent for all feedstock mixtures (Riffaldi et al., 1986).

Compost maturity testing during a 140-d process indicated 2 mo as suitable to attain desired stability in the product (Riffaldi et al., 1986). Parameters of total and hydrolyzed N, C/N ratio, humified C, and optical density indicated maturity after 30 d. Analyses of CEC, organic C, nitric and ammoniacal N, cellulose, hemicellulose, lignin, phenols, and C_{HA} indicated maturity after the 2-mo period. Comparison with biological parameters confirmed 60 d as the start of the maturation period during composting. Ammonia-producing and proteolytic bacteria completed metabolism within 30 d. Nitrification took place during mid-process, after the thermophilic stage. Cellulolytic bacteria were most active after 30 to 60 d and later subsided to mainly fungal and actinomycete activity. Chemical differences in feedstock materials of different composts caused variability in analytical values of the finished composts. More than one parameter must be considered to define the degree of stability in a given compost and definite evaluation can only be from field experiments.

Inoko et al. (1979) proposed a guideline of appropriate ranges for a mature MSW compost. The C/N ratio should be <20:1, total N should be above 2% (oven-dry basis), and the ratio of C in reducing sugars to total C should be below 35%. They recorded decreases in total C, hemicellulose and cellulose, and increases in total N, crude ash, and lignin during maturation of city refuse compost. The authors reported that maturity parameters varied according to feedstock. Cellulostic composts were considered mature with a C/N ratio below 20:1, higher than 2% total N (oven-dry basis), and a reducing sugar ratio below 35%. Cation-exchange capacity was dependent on feedstock and comparison with values gathered during composting. Compost from MSW was reported as similar for cellulostic materials and paper chromatography also was used. These findings were confirmed by a later study monitoring maturity of MSW (Harada et al., 1981); the authors suggesting that compost can be applied to soil after 3 to 4 wk, provided critical values reported in Inoko et al. (1979) are reached.

Maturity tests for agricultural composts appear inconsistent, such as lack of correction of C/N ratio. Despite inconsistencies, Godden and Penninck (1986) suggested that the phosphatase activity index and plotting with time the evolution of different N forms are useful as a basis for a rational composting test. A high and stable phosphatase activity value was correlated with maturity. A low, increasing value denoted composting as near completion while a low but stable or decreasing value denoted poor conditions. They found that the ratio NO_3^-/NH_4^+ in mature materials favors the oxidized form. The authors also suggested using the total mass (rather than the concentration) of mineral and organic forms of N. The C/N, ash percentage, and alkaline phosphatase activity proved potentially useful for estimating maturity.

Organic matter fractionization, degree of decomposition, and total organic matter were analyzed for 43 composts differing in origin and maturity (More & Sana, 1987). A 7-d complementary mineralization rate (CMR) and differential chemical parameters (total organic matter, oxidable C, humic fractionation, colorimetric index of humic extract, decomposition degree, and nonhydrolyzable N) were determined. A statistical method was established to determine the stage of

Table 6–1. C/N ratios of different kinds of well-mature compost and their corresponding raw material (Chanyasak et al., 1983a).

	Raw materials		Composts	
	C_{org}/N_{org} water extracts	C/N of solid samples	C_{org}/N_{org} water extracts	C/N of solid samples
Sewage sludge	5.6	8.7	6.01	11.2
Sewage sludge + sawdust	7.2	21.5	4.34	17.9
Sewage sludge + rice hulls	6.1	14.9	5.50	14.2
Garbage + bark	29.5	21.6	6.36	16.1
Garbage	13.9	16.0	5.72	15.8
Municipal refuse	26.2	20.7	6.85	14.9
Cow dung	12.0	22.0	5.16	11.3
Cow dung + pig manure + straw	-	-	6.15	12.7
Chicken manure	8.6	5.3	5.68	8.2
Leaves	3.0	33.4	5.00	12.1

maturity from characteristics of the organic fractions. Twenty six of the samples were classified with certainty as to their maturity (fresh, semi-mature, or mature); specific analytical problems or conflicting results inhibited classification of the remaining 17.

Carbon to Nitrogen Ratio

Traditionally C/N ratio has been used as an indication of the potential for N mineralization or immobilization transformations. General accepted numbers for soils are: below 20:1 mineralization should occur (organic N decomposes to NH_4^+), and above 30:1 immobilization should occur (NH_4^+ or NO_3^- microbially changed into organic N). Table 6–1 presents the C/N ratios of a number of feedstocks and corresponding composts (Chanyasak et al., 1983a). Jimenez and Garcia (1989) has compared initial to final C/N ratios in an attempt to relate them to maturity (Table 6–2). The C/N ratio of an MSW compost was 30:1 for the raw material and decreased to 13:1 with composting (Charpentier & Vassout, 1985).

Table 6–2. C/N ratios and compost maturity (Jimenez & Garcia, 1989).

Composting days	Initial C/N	Final C/N	Final C/N / Initial C/N	References
63	31.1	19.0	0.61	Parra, 1962
63	29.0	18.0	0.62	Parra, 1962
40	27.0	14.0	0.52	Kehren, 1967
30–180	23.2	16.3	0.70	AGHTM, 1975
180–360	23.2	13.4	0.58	AGHTM, 1975
365	23.2	11.9	0.51	AGHTM, 1975
120	30.3	22.6	0.75	Juste, 1980
240	30.3	22.6	0.75	Juste, 1980
Not indicated	24.0	15.0	0.63	De Bertoldi and Zucconi, 1980
Not indicated	20.7	14.9	0.72	Chanyasak and Kubota, 1981
120	21.5	16.1	0.75	Chanyasak et al., 1982
30	22.3	19.0	0.85	De Bertoldi et al., 1982b
140	34.4	16.7	0.49	Clairon et al. 1982
70	33.0	18.0	0.55	Levasseur and Saul, 1982
90	23.6	15.9	0.67	Lavoux and Souchon, 1983

Municipal waste composts from various cities differed in their components and length of stabilization periods (Inoko et al., 1979). Total C ranged from 27 to 41%, C/N ratios 13 to 31:1, ash 16 to 47%, and hot-water soluble organic matter 3 to 16%. Variations in reducing sugars in the hydrolysate were a hemicellulose range of 3 to 16%, cellulose range of 8 to 35%, and a range in the rates of C in reducing sugars to total C of 17 to 49%. Lignin contents ranged 12 to 28%. Total N varied least at 1.2 to 2.7%. Samples with high C/N and high hemicellulose and cellulose contents were generally low in N and lignin, high in hot-water soluble organic matter. The greatest correlation (0.90) was between C/N and the rates of C in reducing sugars to total C. Observed trends (no statistics) during decomposition and curing, were decreasing C contents and C/N ratios, and linear increases in N contents. Hemicellulose and cellulose contents increased before rapidly decreasing, while lignin increased initially, decreased till Day 15, then gradually increased till the end of the study.

Total C, C/N ratio, the contents of cellulose and hemicellulose, and the ratio of the C in reducing sugar to the total C decreased during MSW composting (Harada et al., 1981). The percentages of total N, ash and lignin increased. The CEC increased to 70 to 80 mol_c kg^{-1} within 5 to 8 wk, then became constant. All constituents became nearly constant after 5 wk of curing with aeration and turning. Organic composition and CEC of compost piled without turning for 12 wk were similar to those of the pile receiving aeration and turning at 2 to 3 wk. Wong (1985) also found that during storage, organic C content and C/N ratio of an MSW compost decreased and total N content increased.

The C/N ratio of a eucalyptus compost shot up from 69 (raw material) to 126 and then decreased to 22 (product) in 432 d (Hardy & Sivasithamparam, 1989). This was compared with a C/N of 52 for a Nurseryman's mix. Except for S and B, elemental concentrations were higher in the compost product than in the nursery mix. Composting decreased electrical conductivity to below the nursery mix. Cation-exchange capacity did not change; it remained higher than the nursery mix.

le Bozec and Resse (1987) concluded that MSW composts reached maturity at a C/N ratio below 20:1. They then assigned appropriate composting time frames in accordance with method and season: 20 wk in summer and 22 wk in winter for windrows; 17 wk in summer and 15 wk in winter for static aerated piles; 12 wk in summer and 8 wk in winter for aerated piles with periodic turning. Mature composts may reach 40 to 60°C and pH tends toward acidity in the first month before becoming basic at 8 to 8.5. Fresh compost should be stacked in windrows 3 m high on a concrete pad and turned twice in the first 6 to 8 wk. Piles should be restocked between the 10th and 12th week to 3 to 4 m high. Watering during stacking should be done in summer to attain moisture between 50 and 55%.

Sugahara and Inoko (1981) suggested that simple determination of C/N is not an accurate test for maturity if an additional N source is added. Chanyasak et al. (1983a,b) tried to relate compost maturity to the ratio of organic C to organic N in water extracts. More specifically, direct determination of the C/N_{org} ratio in water extracts was 5.88 for mature composts. The ratio of total organic C to total organic N (C/N_{total}) in water extracts decreased and attained near linear relation with the organic N to total N ratio (N_{org}/N_{total}), and the C/N_{org} ratio became constant at maturity. The findings agreed with the authors' previous conclusion (Chanyasak & Kubota, 1981) that mature composts have a C/N_{org} ratio in water extracts of between 5 and 6. Carbon to nitrogen ratios in water extracts changed more significantly than C/N ratios of solid samples. The relationship between N_{org}/N_{total} and C/N_{total} (water extract) was expressed as $N_{org}/N_{total} = (1/5.88)(C/N_{total})$, leading to $C/N_{org} = 5.88$. Water extracts had definite correlation while the relation between C/N and N_{org}/N_{total} in solid state did not.

Cation-Exchange Capacity

Cation-exchange capacity is a routine measure of the potential of a soil as a mechanism to hold cations (such as trace metals or NH_4^+) through electrostatic attraction. It is related to the amount of organic matter and the proportion of clay; both of which have negatively charged sites that attract the positive charge on cations. Thus, high CEC is a desirable aspect for the soil.

Cation-exchange capacity was determined by saturating 25 different compost samples with acid resin and removing H^+ ions (Estrada et al., 1987). Values were assayed with or without other chemical parameters of organic fraction, resulting in a useful maturity index. The CEC values ranged from 19.7 to 66.7 mol_c kg^{-1}. The organic fractions had the highest CEC value (50–150 mol_c kg^{-1}) and generally underwent significant change with time, compared with the mineral fractions (10 mol_c kg^{-1}) that did not vary much during composting. The ratio of CEC to total organic matter (CEC/TOM) values correlated with the index of maturity obtained from other chemical parameters: degree of decomposition, humic and fulvic acids (extraction and polymerization rate), and colorimetric index of the humic extract. The relationship among the parameters was determined with the CMR, a measure of CO_2–C emission. The method was reported as reliable with good reproducibility. A more mature compost has a lower CMR (More & Sana, 1987).

A very high correlation ($r = 0.951$) was reported between organic matter and organic C, as both decreased during 140 d of composting (Riffaldi et al., 1986). Organic C decline leveled off after 2 mo, however, the authors did not define this leveling off as a quantitative measure of compost stabilization. The CEC values tended to increase by 50%. Total N increased by 30% after 30 d of composting and then leveled out.

Jacas et al. (1987) reported a high correlation ($r = 0.66$–0.94) between compost age and CEC, despite different feedstocks. The authors recommended CEC testing as a routine maturity test for its precision, simplicity, low cost, and speed (testing took approximately 5 h). Saturation time was reduced for simplification. An assay was made of different samples unmilled, milled, and pulverized to 0.75 and 0.12 mm. The CEC increased in all four treatments. The authors considered pulverization to 0.75 mm sufficient treatment. To use CEC for determining compost maturity, the initial CEC must be known. The authors obtained target CEC values by grouping composts of various feedstocks according to similarity of slope: 0.40 slope for pine bark with or without sewage sludge; 0.26 slope for MSW and pulp and paper sludge with sewage sludge. A 10^{-3} slope was attained for the bark–sewage composts intended for potting mediums. Composts of the same basic feedstock reached similar final CEC values. Bark composts reached 150 mol_c kg^{-1} and MSW composts reached 80 mol_c kg^{-1}.

Nitrification

Nitrification is the oxidation of NH_4^+ to NO_2^-, and NO_2^- to NO_3^- by microorganisms. This occurs when the competition for N is low enough that excess is available. Finstein and Miller (1985) defined compost maturity by nitrification, reporting that compost is ready for use when NO_2^- and /or NO_3^- appear. Riffaldi et al. (1986) considered decreasing then stabilizing NH_4^+ and increasing then stabilizing NO_3^- suitable for evaluation of the degree of compost maturity. Hydrolysable N content decreased by 13% in up to 30 d of composting before becoming constant. The C/N ratio (indicates humification of organic matter) reacted similarly, but exhibited a slow decrease after 30 d. The NH_4^+ portion of N fell after an increase and ultimately accounted for 15% of the initial N. Value for NO_3^-–N had doubled by the end of the experiment.

Others suggest the occurrence other forms of N as useful indicators of maturity. Harada et al. (1981) considered total N as a useful maturity indicator. Raw eucalyptus bark contained more total available N than a pine bark nursery mix, and the amount increased after composting (Hardy & Sivasithamparam, 1989). Riffaldi et al. (1986) observed that total N content, expressed as percentage of dry matter and referred to an unchanging base (ash) remained constant during composting. It was considered more useful from a product use standpoint as reported on a dry matter basis. The two maturity parameters of hydrolysable N and C/N ratio demonstrated a high correlation and acted as useful indicators.

pH

The pH of eucalyptus compost increased from 4.0 to 7.5 in 10 to 12 d, fell to 6.2 to within 65 d, then leveled out (Hardy & Sivasithamparam, 1989). The pH values of two MSW compost piles, one covered and one exposed, were constant for 3 mo, increased by one unit during the fourth month and fell the beginning of the fifth (Charpentier & Vassout, 1985). Leachate analysis of cranberry (*Vaccinium* sp.) processing wastes during composting showed that an addition of lime provided adequate buffer effect (Blanc & O'Shaughnessy, 1989). In contrast, a pile without lime addition failed to maintain an optimum pH range. Others dismissed compost pH as a suitable parameter to assess compost maturity because it cannot be described as a monotonic function (de Nobili & Petrussi, 1988).

Organic Matter and Humification

Ultimately, the readily decomposable organics should tend towards humic substances. Both organic matter and humification has been suggested as indicators of maturity. Organic matter content cannot be used as a maturity test because its initial value or its decrease during time is dependent on the feedstock composition (de Nobili & Petrussi, 1988). Not surprising, Estrada et al. (1987) found that total organic matter decreased during composting. Charpentier and Vassout (1985) also found the amount of total organic matter decreased, from 50 to 60% at the start of composting to 30% 3 mo later. This trend was similar for two compost piles, one covered and one exposed. Wong (1985) found that organic C content reduced from 40 to 30% after 16 wk.

One compost has stabilized, increases in CEC, total N, C_{HA}, lignin, and methoxyl groups are a result of decreases in organic matter content (simple mass loss rather than an absolute increase of these parameters; Riffaldi et al., 1986).

Riffaldi et al. (1983) suggested that research into numerous samples at different stages of maturity and a comparison of the humification of different materials in soil and their effects on crop growth is needed to determine the most suitable sources for soil amendments. They found in experiments that humic and fulvic acids extracted from different organic wastes varied in chemical composition, but generally had high N contents and had characteristics of young humic substances, namely a lower degree of carbonization, lower contents of the main acidic groups, and a more complex organic structure than humic compounds taken from soil organic matter.

Humic acid extracts from fresh and aged bark composts differed with increased humic acid yield, total acidity, total N, and carboxyl groups in the older compost (Albrecht et al., 1982). Alkaline phosphatase activity held a constant high value at the end of curing in optimal composting conditions, but never reached a high value with poor conditions (Godden & Penninck, 1986). Humic acid (HA) increased with a decrease in fulvic acid (FA) as MSW compost stabilized (Sugahara & Inoko, 1981). The shape of infrared (IR) spectra of humic acid, a function of oxidation, became similarly featureless during the curing process, and was suggested to indicate maturity.

Calculations for some parameters have been suggested by Roletto et al. (1985). The humification ratio is the percentage of total extracted humic C (C_{ext}) to organic C:

$$\text{humification ratio} = (C_{ext})(100/C_{org})$$

The humification index (HI) is the percentage of humic acid C (C_{HA}) in an organic C.

$$HI = (C_{HA})(100/C_{org})$$

The authors also suggested minimum values for evaluating maturity of compost made from ligno-cellulosic materials as listed in Table 6–3.

Table 6–3. Minimum values for evaluating maturity of compost made from ligno-cellulosic materials (Roletto et al., 1985).

Parameter	Minimum value
C_{HA}/C_{FA}	1.0
Humification ratio, %	7.0
Humification index, %	3.5
Total humic C, %	3.0
Particles with nominal molecular weight ≥10 000%	40

de Nobili and Petrussi (1988) reported a linear decrease with time in the HI during the thermophilic phase. Values dropped after the pile was turned but increased again each time pile temperature reached 65°C. Water extracts of the HI_w decreased hyperbolically with time in the thermophilic stage. Measurement of HI_w was not considered useful for assessing compost maturity; values indicated stabilization while the thermophilic stage was in progress (de Nobili & Petrussi, 1988). The HI was determined for MSW composts at different stages as the ratio of nonhumified (nonphenolic) to humified (phenolic) organic C after extraction with alkaline sodium pyrophosphate.

Classic analysis of humified organic matter by humic acid (HA) and fulvic acid (FA) contents was not satisfactory for determining humification of organic materials (Riffaldi et al., 1986). The authors cited Sugahara and Inoko (1981) who reported that total humus remained constant while HA increased and FA decreased. Riffaldi et al. (1986) examined their procedure and noted that most changes to the humified C ($C_{HA} + C_{FA}$) had occurred within the first 30 d as a 15% increase. The C_{HA} increased considerably up to 60 d and C_{FA} decreased. The C_{HA}/C_{FA} ratio increased by 50% after a 2-mo stabilization period. This indicated an advanced amount of humification. Optical densities increased significantly the first 20 d and continued at a lesser rate. The extinction values of the pyrophosphate extract result correlated with C/N and C_{HA+FA}, demonstrating reliability for maturity testing.

Chemical degradation with K persulphate and ^{13}C-NMR magnetic or resonance spectral characteristics showed differences in humic acids of three MSW composts much better than analytical parameters such as elementary composition, functional group contents and E4/E6 ratios (Gonzalez-Vila & Martin, 1985).

Gel Chromatography

Gel chromatography has shown that peptide-containing compounds in water extracts change to higher molecular weight compounds early during composting with optimal conditions (Chanyasak et al., 1982). Gel chromatograms of water extracts from composting pulp and paper sludge with straw showed an initial increase of molecular weight followed by a decrease and disappearance (Saviozzi et al., 1987). The same trend occurred with C, N, volatile acid, and amino acid contents.

Spectroscopy

Chemical and spectroscopic analyses were used to plot changes in organic constituents of a cattle manure compost (Inbar et al., 1989). Carbon-13 NMR spectra was acquired with Cross-Polarization Magic Angle Spinning (CPMAS), and infrared spectra was measured with a Fourier-Transform InfraRed spectrophotometer (FTIR). The methods recorded a decrease in carbohydrates, resulting in an accumulation of modified lignin, and increases in levels of alkyl C, aromatic C, and carboxyl groups. FTIR absorbance ratios showed distinct linear correlation with parameters such as compost age, CEC, humic content, and C/N ratio, using peaks for aliphatic C–C, aromatic C=C, COO$^-$, and polysaccharides C–O. The polysaccharides and aliphatic C–H decreased during composting while the concentration of the aromatic C=C alkyl C and carboxylate ions increased. Findings were in general agreement with the NMR spectra. Humic content and CEC increased while the C/N ratio decreased during composting. CPMAS ^{13}C-NMR spectra produced a reliable estimate of lignin content as compared to chemical analysis and without extraction.

Chemical, IR and NMR spectroscopy, and molecular characteristics of humus in MSW were reported before and after composting (Giusquiani et al., 1989). Acidic groups and straight aliphatic chains diminish during composting. Humification of compost after its addition to soil can be recorded with the ratio of peaks in the elution curves on Sephadex G-100. Data showed that humic acids in composted material are lower in acidic groups and have higher E_4/E_6 ratios.

Solid-state multinuclear NMR was used to characterize samples of composts from fish wastes (Preston et al., 1986). Data from NMR analysis proved consistent with other chemical, physical, and biological methods for determining compost maturity and quality. Analysis with NMR can provide chemical information on complex, heterogeneous, largely-insoluble samples, in extracted and solid forms.

Microbial Activity Indications

Respiration

As organic materials are decomposed aerobically by microbes, CO_2 is given off. Thus, the rate of CO_2 evolution has been considered as a useful measure of composting activity. Conversely, as rates significantly reduce, compost may be approaching maturity, if other environmental factors are not limiting (N availability, moisture). Peaks and falls in CO_2 evolution during composting coincided with those in microbial activity (Hardy & Sivasithamparam, 1989). Haug (1980) considered a sufficient level of stabilization to occur when the rate of O_2 reduction does not produce adverse results under storage or use, but degradation is not so complete that organics are lost unnecessarily (Willson & Dalmat, 1986). Microbial respiration rate should indicate maturity. Heat, CO_2 production, and O_2 consumption should also indicate maturity because of direct relation to microbial respiration.

Biological oxygen demand (BOD) was determined significant as both an indicator of maturity and a tool for studying process (Stahlschmidt, 1978). A simplified version of the Warburg method of respiration was applied with sludge compost. Measurements were recorded for the change in partial pressure of O_2 in a sealed container filled with compost. A decrease in BOD of 75% occurred within 30 d of aerobic curing. Compared with organic matter content, C/N ratio, the Chatomim test, and a growth test, it was concluded that given optimal composting conditions, maturity is reached when BOD is <50 mg O_2 kg^{-1} at 50% moisture h^{-1} and remains so for some days. Reliability of this method is dependent on control of temperature and homogeneity of samples. Correlation with bioassays was suggested for the purpose of quality control during composting.

Comparisons have been made between composts and soils for further information on respiration techniques as a maturity test. The priming effect, a quantification of the increase in soil respiratory rate (due to mineralization) with increasing doses of compost, was studied as an index for compost maturity (Ribalta et al., 1987). Evolved C, as CO_2, was plotted vs. application. Respirometrical rate-compost dose ratio (RR/CD) was indicated by the regression line slope. Application of fresh compost produced steeper slopes than mature composts and semi-mature composts generated intermediate slopes. Behavior was consistent for three tested soils. To assure consistency, the soils should not be too high in organic matter (<25%), a pH close to neutral, and a small amount of carbonates.

Physical Indications

Physical signs of stability, such as loss of self-heating and general appearance: uniformity, dark color and earthy smell, are most reliable when compared with other parameters.

Smell

Although there has not been quantitative measurements for smell (for obvious reasons of subjectiveness), it seems reasonable that compost that maintains an earthy smell while storage will be mature enough not to cause detrimental effects plants. Conversely, a compost with an obnoxious odor suggests instability.

Temperature

Earlier studies have proposed that temperature trends, or reheating under laboratory conditions, indicate maturity (Harada et al., 1981). Compost was considered mature once the temperature levels off and did not react to turning. The authors found that temperature of an MSW windrow increased with turning after 8 wk of piling, despite other indications of stability within 5 wk. They concluded that the rise resulted from the activity of a small amount of organic matter that produced negligible amounts of change in the overall composition. Temperature was dismissed as a dependable indication of maturity. Also, temperature increases may not occur if conditions are not conducive to composting (i.e., moisture limitations).

SUMMARY

Maturity, as we have defined it—the point at which the compost will not act detrimentally when used as a soil amendment—is a somewhat arbitrary term. Thus, it would seem, tests trying to assess maturity must bear this same degree of capriciousness. It is apparent from the above discussions, that many tests have been proposed for assessing the maturity of composts. For the most part, there exists some degree of correlation between particular tests and the maturity (or stability, or age) of the compost being tested. Thus, many tests can succeed (when used correctly) in their goal of providing guidance to the composter on operational requirements (time, temperature, moisture, or aeration) for producing a mature compost. It is the authors' opinion, however, that there has not yet been devised a test (with corresponding critical values) that will definitely assure that all composts that pass will act as desired in the soil.

Tests differ in their simplicity, duration, and approach. The most direct include smelling (for obnoxious odors) and plant growth or germination tests. All the rest may be termed indirect measures of maturity, since they assess chemical or biological parameters that may correlate with plant effects or odors. The simplest include the physical indications (the earthy smell or temperature/reheating). The more sophisticated include fractionation into categories of organic compounds or respiration using a variety of methodologies. There appears to be little agreement among scientists which is the best test(s), and thus, this area of compost science will continue to be a strong field of research.

REFERENCES

Association Génerale des Hygienistes et Techniens Municipaux. 1975. Residus urbains. Technique et Documentation. AGHTM, Paris, France.

Albrecht, M.L., M.E. Watson, and H.K. Tayama. 1982. Chemical characteristics of composted hardwood bark as they relate to plant nutrition. J. Am. Soc. Hortic. Sci. 107(6):1081–1084.

Blanc, F.C., and J.C. O'Shaughnessy. 1989. Static pile composting of cranberry receiving wastes and processing residues. p. 569–578. In Proc. of the 43rd Purdue Industrial Waste Conf., Purdue Univ., Lafayette, IN. 10–12 May 1988. Lewis Publ., Chelsea, MI.

Chanyasak, V., M. Hirai, and H. Kubota. 1982. Changes of chemical components and nitrogen transformation in water extracts during composting of garbage. J. Ferment. Technol. 60(5):439–446.

Chanyasak, V., A. Katayama, M.F. Hirai, S. Mori, and H. Kubota. 1983a. Effects of compost maturity on growth of komatsuna (*Brassica Rapa* var. *pervidis*) in Neubauer's pot: I. Comparison of growth in compost treatments with that in inorganic nutrient treatments as controls. Soil Sci. Plant Nutr. 29(3):239–250.

Chanyasak, V., A. Katayama, M.F. Hirai, S. Mori, and H. Kubota. 1983b. Effects of compost maturity on growth of komatsuna (*Brassica Rapa* var. *pervidis*) in Neubauer's pot: II. Growth inhibitory factors and assessment of degree of maturity by org.-C/org.-N ratio of water extract. Soil Sci. Plant Nutr. 29(3):251–259.

Chanyasak, V., and H. Kubota. 1981. Carbon/organic nitrogen ratio in water extract as a measure of composting degradation. J. Ferment. Technol. 59:215–219.

Charpentier, S., and F. Vassout. 1985. Soluble salt concentrations and chemical equilibria in water extracts from town refuse compost during composting period. Acta Hortic. 172:87–93.

Clairon, M., C. Zinsou, and D. Nagou. 1982. Etude des possibilités d'utilisation agronomique des composts d'ordures ménagères en milieu tropical: I. Compostage des ordures ménagères. Agronomie. 2:295–300.

De Bertoldi, M., G. Vasllini, A. Pera, and F. Zuconni. 1982. Comparison of three windrow compost systems. Biocycle. 23:45–49.

De Bertoldi, M., and F. Zuconni. 1980. Microbiologia della transformazione dei rifiuti solidi urbani in compost e lor utilizzazione in agricoltura. Ingegneria ambientale. 9:209–216.

de Nobili, M., and F. Petrussi. 1988. Humification index (HI) as evaluation of the stabilization degree during composting. J. Ferment. Technol. 66(5):577–583.

Estrada, J., J. Sana, R.M. Cequiel, and R. Cruanas. 1987. Application of a new method for CEC determination as a compost maturity index. p. 334–340. In M. de Bertoldi et al. (ed.) Compost: Production, quality and use. Udine, Italy. 17–19 Apr. 1986. Elsevier, London.

Finstein, M.S., and F.C. Miller. 1985. Principles of composting leading to maximization of decomposition rate, odor control, and cost effectiveness. p. 13–26. In J.K.R. Gasser (ed.) Composting of agricultural and other wastes. Seminar by the CEC, Brasenose College, Oxford. 19–20 Mar. 1984. Elsevier, London.

Giusquiani, P.L., M. Patumi, and M. Businelli. 1989. Chemical composition of fresh and composted urban waste. Plant Soil 116(2):278–282.

Godden, B., and M.J. Penninck. 1986. On the use of biological and chemical indexes for determining agricultural compost maturity: Extension to the field scale. Agric. Wastes 15:169–178.

Gonzalez-Vila, F.J., and F. Martin. 1985. Chemical structural characteristics of humic acids extracted from composted municipal refuse. Agric. Ecosyst. Environ. 14(3/4):267–278.

Harada, Y., A. Inoko, M. Tadaki, and T. Izawa. 1981. Maturing process of city refuse compost during piling. Soil Sci. Plant Nutr. 27(3):357–364.

Hardy, G.E.S.J., and K. Sivasithamparam. 1989. Microbial, chemical and physical changes during composting of a eucalyptus (*Eucalyptus calophylla* and *Eucalyptus diversicolor*) bark mix. Biol. Fertil. Soils. 8:260–270.

Haug, R.T. 1980. Compost engineering. Principles and practice. Ann Arbor Sci. Publ., Ann Arbor, MI.

Henry, C.L. (ed.) 1991. Technical information on the use of organic materials as soil amendments. Washington State Dep. of Ecology and Solid Waste Compost Council, Alexandria, VA.

Hoitink, H.A.J., and P.J. Fahy. 1986. Basis for the control of soilborne plant pathogens with composts. Annu. Rev. Phytopathol.. 24:93–114.

Inbar, Y., Y. Chen, and Y. Hadar. 1989. Solid-state carbon-13 nuclear magnetic resonance and infrared spectroscopy of composted organic matter. Soil Sci. Soc. Am. J. 53:1695–1701.

Inoko, A. 1985. Evaluation of maturity of various composted materials. JARQ 19:103–108.

Inoko, A., K. Miyamatsu, K. Sugahara, and Y. Harada. 1979. On some organic constituents of city refuse composts produced in Japan. Soil Sci. Plant Nutr. 25(2):225–234.

Jacas, J., J. Marza, P. Florensa, and M. Soliva. 1987. Cation exchange capacity variation during the composting of different materials. p. 309–320. *In* M. de Bertoldi et al. (ed.) Compost: production, quality, and use. Udine, Italy. 17–19 Apr. 1986. Elsevier, London.

Jimenez, E.I., and V.P. Garcia. 1989. Evaluation of city refuse compost maturity: A review. Biol. Wastes 27(2):115–142.

Juste, C. 1980. Avantages et inconvenients de l'utilisation des composts d'ordures ménagères comme amendement organique des sols ou supports de culture. *In* Int. Conf. on Compost, Madrid Spain. 22–26 Jan. Min. Obras Públicas.

Juste, C., P. Solda, and M. Lineres. 1987. Factors influencing the agronomic value of city refuse composts. p. 388–398. *In* M. de Bertoldi et al. (ed.) Compost: Production, quality and use. Udine, Italy. 17–19 Apr. 1986. Elsevier, London.

Kehren, L. 1967. Le compostage des ordures ménagè dans les pays chauds. Tech. Sci. Munic. 62:211–216.

Lavoux, T., and C. Souchon. 1983. Le compostage. p. 65–83. *In* Optimisation Energetique et Ecologique de Quelques Filiéres de Valorisation des Dechets. Min. De. l'Environnement, Paris.

le Bozec, A., and A. Resse. 1987. Experimentation of three curing and maturing processes of fine urban fresh compost on open areas. p. 78–96. *In* M. de Bertoldi et al. (ed.) Compost: Production, quality and use. Udine, Italy. 17–19 Apr. 1986. Elsevier, London.

Levasseur, J.P., and W.B. Saul. 1982. Composting of urban solid waste. p. 81–85. *In* Proc. of a Conf. on the Practical Implications of the Reuse of Solid Waste, London. 11–12 Nov. 1981. Thomas Telford, London.

Mathur, S.P., J.Y. Daigle, M. Lévesque, and H. Dinel. 1986. The feasibility of preparing high quality composts from fish scrap and peat with seaweeds or crab scrap. Biol. Agric. Hortic. 4:27–38.

More, J.C., and J. Sana. 1987. Criteria of quality of city refuse compost based on the stability of it organic fraction. p. 321–327. *In* M. de Bertoldi et al. (ed.) Compost: Production, quality and use. Udine, Italy. 17–19 Apr. 1986. Elsevier, London.

Parra, J.H. 1962. Fabricación de compost a partir de basuras. Cenicafé (Columbia) 13:51–68.

Preston, C.M., J.A. Ripmeester, S.P. Mathur, and M. Levesque. 1986. Application of solution and solid-state multinuclear NMR to a peat-based composting system for fish and crab scrap. Can. J. Spectrosc. 31(3):63–69.

Ribalta, R., J.C. More, and J. Sana. 1987. The priming effect and the respiratory rate/compost dose ratio as compost ripeness index. p. 328–333. *In* M. de Bertoldi et al. (ed.) Compost: Production, quality and use. Udine, Italy. 17–19 Apr. 1986. Elsevier, London.

Riffaldi, R., R. Levi-Minzi, A. Pera, and M. De Bertoldi. 1986. Evaluation of compost maturity by means of chemical and microbial analyses. Waste Manage. Res. 4(4):387–396.

Riffaldi, R., R. Levi-Minzi, and A. Saviozzi. 1983. Humic fractions of organic wastes. Agric. Ecosyst. Environ. 10(4):353–359.

Roletto, E., B. Barberis, M. Consiglio, and R. Jodice. 1985. Chemical parameters for evaluating compost maturity. BioCycle 26(2):46–47.

Saviozzi, A., R. Riffaldi, and R. Levi-Minzi. 1987. Compost maturity by water extract analyses. p. 359–367. *In* M. de Bertoldi et al. (ed.) Compost: Production, quality and use. Udine, Italy. 17–19 Apr. 1986. Elsevier, London.

Stahlschmidt, V. 1978. The biological oxygen demand by composting. *In* 3rd Brasilian and 1st Pan American Public Cleaning Congress, Sao Paulo, Brazil. 22–25 Aug. 1978.

Sugahara, K., and A. Inoko, 1981. Composition analysis of humus and characterization of humic acid obtained from city refuse compost. Soil Sci. Plant Nutr. 27:213–224.

Van Soest, P.J. 1963. Use of detergents in the analysis of fibrous feeds: II. A rapid method for determination of fiber and lignin. J. Assoc. Off. Anal. Chem. 49:546–551.

Willson, G.B., and D. Dalmat. 1986. Measuring the degree of compost stability. Biocycle 27(7):34–37.

Wong, M.H. 1985. Phytotoxicity of refuse compost during the process of maturation. Environ. Pollut. Ser. A 37(2):159–174.

Wong, M.H., and L. Chu. 1985. The responses of edible crops treated with extracts of refuse compost of different ages. Agric. Wastes. 14:63–74.

Zucconi, F., A. Pera, M. Forte, and M. de Bertoldi. 1981. Evaluating toxicity of immature compost. BioCycle. March/April:54–57.